HOW
SCIENCE
WORKS

Making up a solution

Steady hand game using a simple electric circuit

Testing action and reaction

Testing the power
of air pressure

HOW
SCIENCE
WORKS

Judith Hann

Hot-air balloon to
show expansion

Testing water hardness

Timing speed
and acceleration

The Reader's Digest Association, Inc.
Pleasantville, New York • Montreal

A READER'S DIGEST BOOK
Designed and edited by Dorling Kindersley Limited, London

Project Editor	John Farndon
Project Art Editor	Geoffrey Manders
Editor	Liza Bruml
Designers	Martyn Foote
	Brian Rust
Picture Research	Diana Morris
	Kate Fox
Production	Helen Creeke
Managing Editor	Carolyn King
Managing Art Editor	Nick Harris

Library of Congress Cataloging-in-Publication Data

How science works / Judith Hann.
p.   cm. – (The Family science library)
Includes index.
ISBN 0-89577-382-1
1.   Scientific recreations.   2.   Science – Experiments – Popular
works.   I. Hann, Judith   II. Series
Q164.H85 1991
507.8 – dc20   90-26457

Revised Edition ISBN 0-89577-909-0

Printed in the USA

Contents

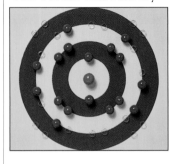

The World of Matter

Energy, Force, & Motion

Light &
Sound

Air &
Water

Electricity &
Magnetism

Electronics &
Computers

INTRODUCTION

It has never been more important to understand the scientific world around us. Our newspapers and televisions are now full of stories showing concern over global warming, transportation technology, the ethical problems of gene therapy, surrogate mothers, and many other significant issues.

To understand all these issues, we need scientific knowledge. That is what this book is all about.

The following pages lead you through the basic principles of chemistry and physics in a simple and colorful way. This is meant to be a practical book, to use around the house, because the experiments that illustrate the main scientific principles use everyday objects in a "home laboratory." It is planned to provide accessible science for the whole family. Many of the experiments can be safely carried out by young children. Parental supervision is recommended throughout. Experiments involving heat and potential hazards *require* adult supervision, and carry a warning to that effect.

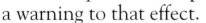

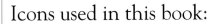

You can use the book by starting at the beginning and proceeding through all the experiments to the end. But it is written in sections that work on their own, if you prefer to dip into different subjects at different times. I hope that after using the book you will realize that science is not the remote, dull subject it can often seem when studied at school. Science makes sense of the world around us, and the best way to learn is to experiment and observe.

I hope that the experiments chosen in the book will stimulate a deeper interest and understanding of scientific subjects. I hope it will inspire more young people to become scientists later in life, while helping adults to understand the scientific principles behind everyday events in their lives, like seeing a rainbow, riding a bicycle, or even making coffee.

Judith Hann

Icons used in this book:

 Adult supervision is required.

 Adult supervision is advised.

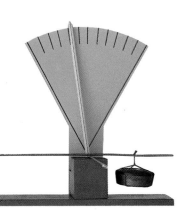

The home laboratory

YOU DO NOT NEED a full–scale laboratory and a closet full of expensive, special equipment to try scientific experiments. Most of the experiments in this book can be carried out with simple, everyday materials that can be found around the home — especially in the kitchen — or acquired very easily. On this and the following three pages, we show some of the items you might find useful in your home laboratory. If you do not have the exact item, improvise.

■ Useful items

The average home is full of items that can be used in all kinds of scientific experiments. Put a box aside for miscellaneous items like these, and drop them in whenever you come across them. Things such as corks, balloons, and drinking straws can be particularly useful. So too can old metal items. Try to find old lenses for experiments with light.

Mirror

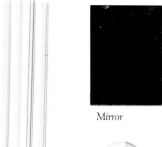

Drinking straws

Safety apparel

Personal safety must always be considered when performing experiments. Wear safety goggles as needed. When working with heaters, tie back long hair and don't wear loose-fitting, long-sleeved clothing.

Magnets

Fuse wire

Old keys

Nuts, bolts, and screws

Balloons, especially large ones

Needles and pins

Lenses

■ Measures

Accurate measurements are needed in many everyday activities as well as scientific experiments. In many kitchens, for instance, you will find measures for volume (measuring jugs), mass (scales and weights), and temperature (cooking thermometers). Before using any thermometer, though, make sure that it covers the right temperature range for the experiment. A garden thermometer, for example, will burst if you try to use it for boiling liquids. Most scientists now use the International System (SI) of measures, with meters for length, kilograms for mass, and seconds for time. But if your measures show other units, use them. Where possible, we have listed appropriate conversions.

Kitchen weights

Spring balance

Thermometer

Plastic measuring cylinder

Kitchen scales

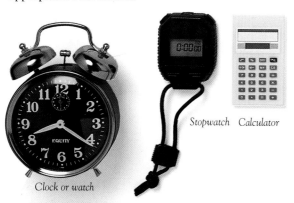

Stopwatch   *Calculator*

Clock or watch

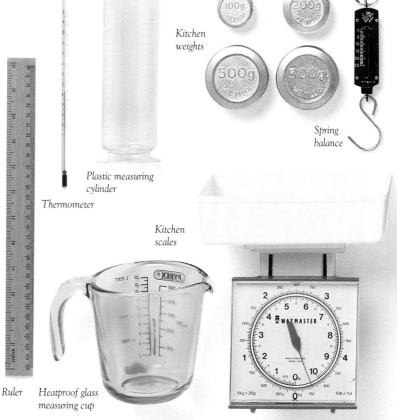

Ruler   *Heatproof glass measuring cup*

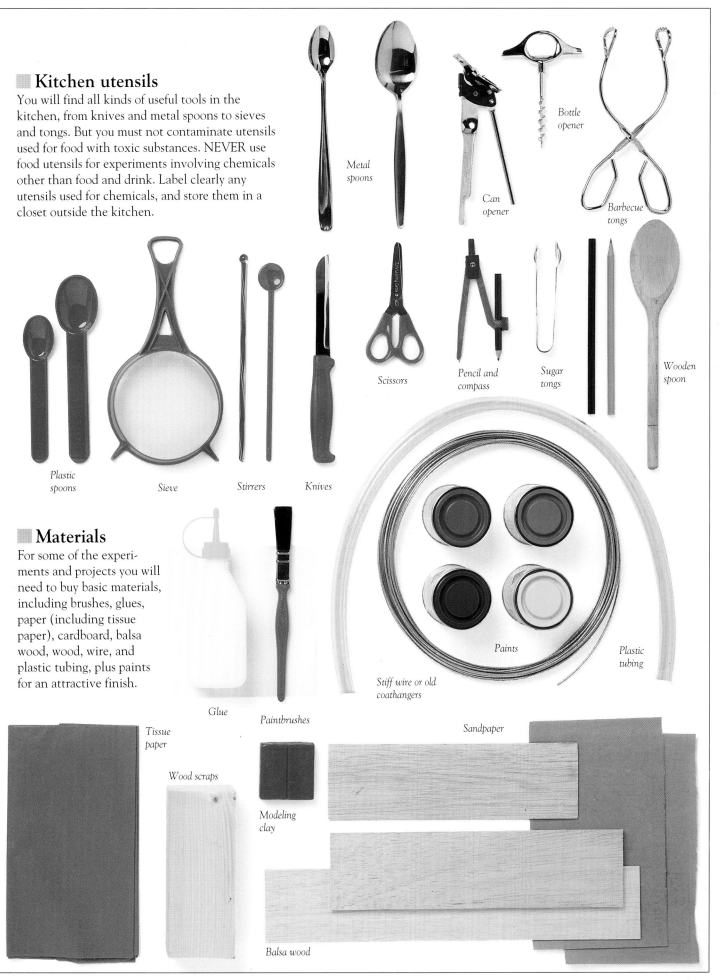

■ Kitchen utensils

You will find all kinds of useful tools in the kitchen, from knives and metal spoons to sieves and tongs. But you must not contaminate utensils used for food with toxic substances. NEVER use food utensils for experiments involving chemicals other than food and drink. Label clearly any utensils used for chemicals, and store them in a closet outside the kitchen.

Metal spoons

Can opener

Bottle opener

Barbecue tongs

Plastic spoons

Sieve

Stirrers

Knives

Scissors

Pencil and compass

Sugar tongs

Wooden spoon

■ Materials

For some of the experiments and projects you will need to buy basic materials, including brushes, glues, paper (including tissue paper), cardboard, balsa wood, wood, wire, and plastic tubing, plus paints for an attractive finish.

Glue

Paintbrushes

Stiff wire or old coathangers

Paints

Plastic tubing

Tissue paper

Sandpaper

Wood scraps

Modeling clay

Balsa wood

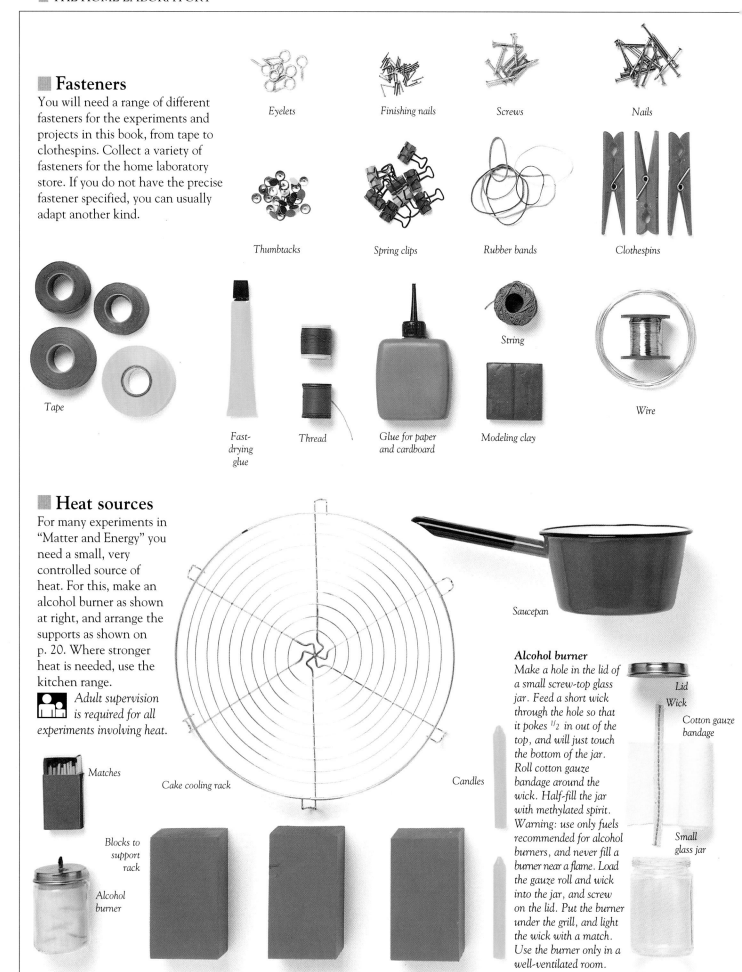

■ Fasteners

You will need a range of different fasteners for the experiments and projects in this book, from tape to clothespins. Collect a variety of fasteners for the home laboratory store. If you do not have the precise fastener specified, you can usually adapt another kind.

Eyelets

Finishing nails

Screws

Nails

Thumbtacks

Spring clips

Rubber bands

Clothespins

Tape

Fast-drying glue

Thread

Glue for paper and cardboard

Modeling clay

String

Wire

■ Heat sources

For many experiments in "Matter and Energy" you need a small, very controlled source of heat. For this, make an alcohol burner as shown at right, and arrange the supports as shown on p. 20. Where stronger heat is needed, use the kitchen range.

Adult supervision is required for all experiments involving heat.

Matches

Cake cooling rack

Candles

Blocks to support rack

Alcohol burner

Saucepan

Alcohol burner

Make a hole in the lid of a small screw-top glass jar. Feed a short wick through the hole so that it pokes $1/2$ in out of the top, and will just touch the bottom of the jar. Roll cotton gauze bandage around the wick. Half-fill the jar with methylated spirit. Warning: use only fuels recommended for alcohol burners, and never fill a burner near a flame. Load the gauze roll and wick into the jar, and screw on the lid. Put the burner under the grill, and light the wick with a match. Use the burner only in a well-ventilated room.

Lid

Wick

Cotton gauze bandage

Small glass jar

■ Tools

A good craft knife is almost essential for many experiments. For the more complex projects, you need basic woodworking tools.

Adult supervision is required for experiments involving sharp cutting tools, drills, and hammers.

■ Electrics

Try to build up a collection of wires, clips, bulbs, and batteries (old and new) for the electrical experiments.

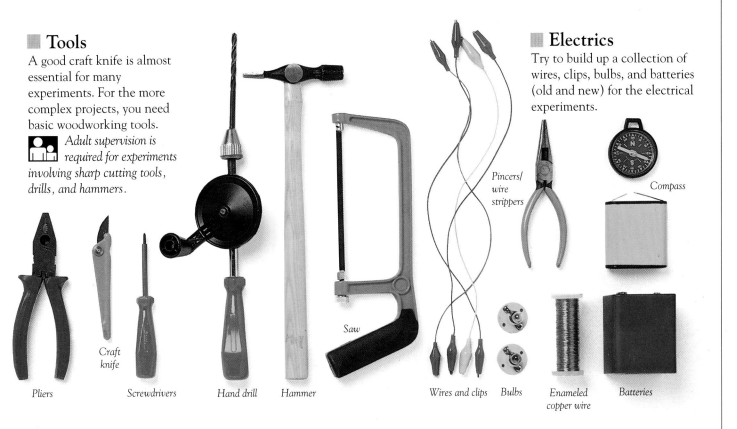

Pincers/ wire strippers

Compass

Pliers

Craft knife

Screwdrivers

Hand drill

Hammer

Saw

Wires and clips

Bulbs

Enameled copper wire

Batteries

■ Containers

You will need a good collection of containers of different types. Old jelly jars are especially useful. So too are used yogurt and other food cartons. Be sure to label contents clearly. NEVER put chemicals in containers in which they could be mistaken for food. It is worth buying a small test tube for heating small quantities. Use this only over a gentle heat.

Small screw-top jars

Plastic cups

Mugs and cups — NEVER use for chemicals

Test tube

Old cans

Used plastic food containers

Old jelly jars

Storage jars

Spice dishes

Heatproof glass beakers

Plastic soft-drink bottles

Large bowls

Old saucers

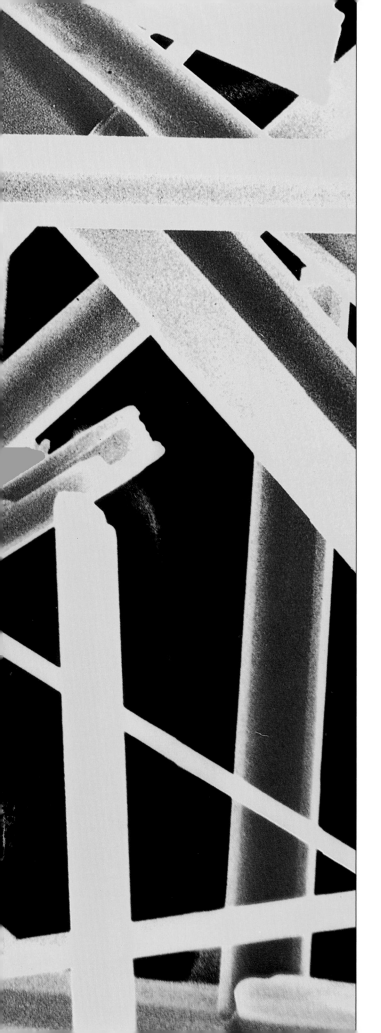

THE
WORLD OF
Matter

WHAT ARE THINGS really made of?
Why are substances like they are?
Can they be changed? These are the
kinds of questions asked in chemistry,
which is the study of all the
substances that make up our world,
living and nonliving.

*Magnified many times in an electron
micrograph, the crystals of caffeine
from coffee grains look like needles.*

13

WHAT IS MATTER?

WHEN SCIENTISTS TALK about "matter," they mean every substance in the universe, from the tiniest speck of dust to the largest star. Indeed, matter is everything that is not simply empty space, or, rather, matter is anything that takes up space. Matter is often thought of as something solid, but it can also be either a liquid or a gas.

This model of the atom and its shells of orbiting electrons (pp. 26-27) is based on the theory of matter suggested by the Danish physicist Neils Bohr in 1913 and still used by chemists today.

People have wondered about the true nature of matter for thousands of years. In the days of Ancient Greece, over 2,000 years ago, many scientists, or "natural philosophers" as they were called, thought that all matter could be divided into four basic types or elements — earth, water, air, and fire. This sounds odd, but it is not so different from our modern division of matter into solids, liquids, and gases.

Each element, they thought, had its own natural resting place, with earth at or below the ground, water and air above, and fire at the top surrounding the Earth's atmosphere. Every substance tried to return to its natural resting place, they argued, which was why stones fell when you dropped them and flames leaped upward.

Thinkers such as Aristotle (384–322 B.C.) also believed matter was really made of one continuous substance such as some fluid or gas — so it is always possible to chop a lump of matter into smaller pieces.

However, there were other

This experiment (p. 35) shows physical and chemical properties of different liquids.

Greek thinkers, such as Democritus (c. 460–400 B.C.), who argued that matter was made of clumps of tiny particles with empty space in between, just as scientists believe today.

These particles were the smallest possible pieces of matter and could never be chopped up, which is why they were called "atoms," the Greek word for "uncuttable." But it was Aristotle's view that seemed more convincing at the time, and for more than 2,000 years, the idea of atoms was forgotten.

■ The alchemists

In the meantime, many Greek and Arab philosophers studied the moon and the stars and argued that there must be a fifth natural element, called "ether," that filled the heavens. Others began to study a particular kind of chemistry called "alchemy," which flourished not only among the Arabs but also in medieval Europe. Alchemists experimented with matter in order to learn how to change it. Soon they began a long and fruitless quest for the "philosopher's stone," the substance that would change ordinary metal into gold.

Alchemists were as much interested in magic and superstition as in science, and

True solids — that is, most solids but glass — are made of crystals; many can be grown in solution (p. 34).

they are sometimes thought of as wizened wizards stirring strange potions rather than serious scientists. But alchemy attracted some powerful minds, and they made some important discoveries, such as nitric and sulfuric acids, and how to make drugs from herbs. Indeed, the word "chemistry" comes from the Arabic *al quemia*, or alchemy.

■ Elements and compounds

It was not until the 17th century that the alchemists' view of the world and Aristotle's four elements (earth, water, air, and fire) were really challenged. Then the Irish scientist Robert Boyle (1627–91) suggested the idea of basic pure chemicals or elements which could be combined to make particular "compounds." Each of these elements, Boyle argued, had its own characteristics and could exist as a solid, liquid, or gas. Boyle also urged scientists to accept the old idea of atoms again.

Soon many experiments showed that Boyle was right. Joseph Priestley (1733–1804) and the brilliant French scientist Antoine Lavoisier (1743–94), for example, proved that air, one of Aristotle's basic elements, was actually a mixture of different gases, including oxygen and nitrogen (pp. 118-119). Lavoisier also showed that another of Aristotle's basic elements, water,

Like many physical properties, the viscosity (stickiness) of a liquid depends on the structure of its molecules. Oil, for instance, has very big, tangled molecules and is very sticky or viscous. This test for viscosity is explained on pp. 22-23.

All these substances are normally liquids, but once they are hot enough they turn into gases (p. 23).

was a mixture of hydrogen and oxygen (p. 28).

Not long afterward, in 1808, English chemist John Dalton (1766–1844) put forward his atomic theory. He suggested that all the atoms of an element are identical — but different from the atoms of every other element. He also argued that compounds were formed by the joining of an atom of one element with an atom of another. Water, he thought, was made when a hydrogen atom linked up with an oxygen atom.

■ Molecules

Three years later, the Italian physicist Amedio Avogadro (1776–1856) showed that in water, each oxygen atom joined up not with one hydrogen atom as Dalton had said but with two. Moreover, neither hydrogen nor oxygen atoms ever existed alone, but only in pairs. These pairs of atoms are now called "molecules" (p. 28).

As Dalton was working out his ideas, chemists were racing to discover new elements and compounds. Progress was astonishing. Dalton's notebooks from 1803 list just 20 different elements. By 1830, chemists knew of 55. By the end of the 19th century, they knew 80 elements and many more compounds.

For a long while after Dalton put forward his atomic theory, people thought that atoms could never be split up, destroyed, or created. They seemed to be like minute billiard balls, solid and

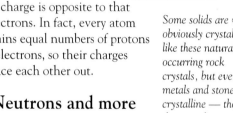

The regular shape of crystals like these of copper sulfate (p. 34) reflects their regular molecular structure.

indestructible. But chemists gradually began to wonder whether this was true.

■ Electrons

It was the British physicist J. J. Thomson (1856–1940) who showed they were right to wonder. In 1897, Thomson's experiments with cathode ray guns (devices something like TV tubes) led him to discover tiny particles, 1,800 times smaller than the smallest atom. These particles were called "electrons" because they have an electrical charge (p. 148).

Thomson believed electrons sit on the atom something like raisins in raisin bread. But further experiments suggested otherwise. In 1911, New Zealand-born Ernest Rutherford realized that at the heart of every atom is a tiny but very dense "nucleus." Around this nucleus, he believed, electrons whirl in a relatively vast cloud, which is largely empty space. Two years later, the Danish physicist Neils Bohr suggested that electrons spin around the nucleus in orbits, like planets around the Sun. These

When substances are mixed together physically like the pigments in ink — but their atoms and molecules remain separate — they can often be separated by simple processes such as chromatography (p. 33).

orbits, Bohr proposed, are arranged in layers or "shells" like the layers of an onion (pp. 26-27).

Then, in 1919, Rutherford split the atom for the first time — it was not indestructible at all. In fact, he smashed the *nucleus* of an atom of nitrogen gas with a stream of alpha particles (helium nuclei). What was left were nuclei of atoms of hydrogen gas, the smallest atoms of all.

It soon became clear that the nucleus of atoms of every element contained hydrogen nuclei, or "protons" as Rutherford called them. Like electrons, protons are electrically charged, but their charge is opposite to that of electrons. In fact, every atom contains equal numbers of protons and electrons, so their charges balance each other out.

■ Neutrons and more

Thirteen years after Rutherford split the atom, the nucleus was shown to contain another kind of particle called a "neutron," as well as protons.

For everyday chemistry, you can still think of the atom as electrons orbiting in shells around a nucleus of protons and neutrons. But physicists have since shown that the story of matter is unimaginably more complicated. Atoms have been smashed many times to reveal the existence of scores of different particles, and these particles are not like billiard balls at all but ghostly packages of energy or "quanta." Scientists are still far from understanding the true nature of matter.

Some solids are very obviously crystalline, like these naturally occurring rock crystals, but even metals and stone are crystalline — though the crystals are usually too small to see.

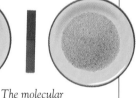

The molecular structure of some substances, such as iron, steel, and cobalt, makes them magnetic (pp. 158-163); most are not. So some mixtures of substances can be separated magnetically (p. 32).

What are chemicals?

THE WORD "CHEMICAL" often conjures up images of acids, steaming liquids in test tubes, and substances with very long names. In fact, every substance in the world is made of chemicals. The sea, the land, plants, buildings, cars, radios, chairs, food, and even your body are all made of chemicals — many as ordinary as salt and water.

Every chemical has its own characteristics. Some are "physical" properties such as color, hardness, texture, and brittleness. The chart below shows some of the most important physical properties, and experiments over the next ten pages explore them in detail. By looking at these properties, we begin to appreciate why different substances behave as they do.

"Chemical" properties are the way chemicals react together, and are not so easy to recognize. Confectioners' sugar and chalk dust, for instance, are physically similar, for they are both fine white powders. But they are different chemically; sugar dissolves in water, and chalk does not. Chemical properties are explored in detail on pages 26-41.

Professor Wildbrain's laboratory
This picture sums up the popular image of the chemistry laboratory — all mysterious steaming liquids, strangely shaped flasks, and imminent explosions. But very few real chemistry experiments ever look like this. In fact, the picture is faked. All the liquids are simply water stained with food coloring, and the fumes are caused by "dry ice," frozen carbon dioxide often used in theaters to create mist effects. It flows down the flask simply because it is heavier than air.

■ Physical properties

Below are listed some important physical properties of chemicals. Look for these properties in substances you find in and around your home. Be as exact as you can — do not just say something is hard, but add that it is harder than one substance and softer than another. See if you can identify groups of substances with similar physical properties.

Solidity (state)	*Is it solid, liquid, or gas? (pp. 20-25)*
Appearance	*What color is it?*
	Is it shiny or dull? (p. 82)
	Is it clear or opaque? (p. 80)
	Is it dark or light?
Texture	*Is it rough or smooth?*
	Is it a powder or a crystal?
Plasticity	*Can it be stretched or bent?*
	Can it be squashed? (p. 74)
	Does it break rather than bend? (p. 17)
Elasticity	*Does it go back to its original shape after stretching? (p. 74)*
Hardness	*How hard is it? (p. 17)*
Density	*How heavy is it? (p. 18)*
Buoyancy	*Does it float? (p. 140)*
Magnetism	*Is it magnetic? (p. 160)*
Conductivity	*Does it conduct electricity? (p. 151)*

EXPERIMENT
Testing for brittleness

Adult supervision is required.
Some substances are hard, but break easily. They are difficult to bend or crush, but crack if hit hard. Such substances are said to be brittle.

YOU WILL NEED
● *household items to test* ● *tweezers to hold small items* ● *hammer* ● *board* ● *safety goggles to protect your eyes*

Hammer test
Place each test substance on a chopping board. Wear the safety goggles to protect your eyes, or cover the substance with a cloth. Give it a short, sharp tap with the hammer. Note what happens when it is struck: does it bend, crack, break, or withstand the blow?

Caution
Don't try this experiment with glass, porcelain, or other materials which might shatter, and send sharp shards flying.

EXPERIMENT
Testing for hardness

YOU WILL NEED
● *collection of household items to test, such as a wooden spoon, plastic bottle, stones, key, modeling clay* ● *plastic pen top*

A simple way of testing a substance's hardness is to scratch it with a plastic pen top. Note how hard you have to press to scratch each substance.

Soft and hard
Hardness cannot be measured in units in the same way as, say, weight or length. However, a German mineralogist named Friedrich Mohs devised a scale against which different substances can be tested. Each standard mineral in the scale can scratch the one below it and can be scratched by the one above it.

Hard rock
Diamond is the hardest substance of all, which is why it is used in industrial cutting equipment. A diamond can only be cut by another diamond.

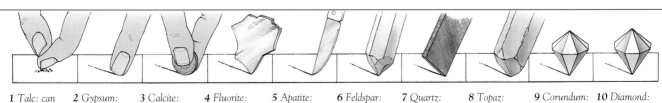

1 *Talc: can be crushed by a fingernail*   **2** *Gypsum: scratched by a fingernail*   **3** *Calcite: scratched by a bronze coin*   **4** *Fluorite: scratched by glass*   **5** *Apatite: scratched by a penknife*   **6** *Feldspar: scratched by quartz*   **7** *Quartz: scratched by hard steel file*   **8** *Topaz: scratched by corundum*   **9** *Corundum: scratched by diamond*   **10** *Diamond: scratched only by itself*

What are chemicals? 2

ONE KIND OF SUBSTANCE you can usually identify easily is metal. Together, metals form one of the most important of all groups of substances. Indeed, all basic substances can be divided into metals and nonmetals (p. 186). There are, of course, many different kinds of metals, but all metals tend to have similar physical properties.

Apart from mercury, which is a silvery liquid at room temperature, most metals are hard and shiny, and clang when you hit them. They are strong, too, and can usually be hammered or molded into almost any shape — which is why they have long been used for making everything from knives to airplanes. They also conduct (carry) both heat and electricity well (pp. 52 and 151).

Although you rarely see their inner structure, all metals are made up of tiny crystals, held together by very tough chemical bonds called "metallic bonds." It is because these bonds are so firm that metals are so strong.

Forging steel
There are many different kinds of steel, but much is made by the oxygen process. New iron is melted in a furnace, impurities are burned out with a blast of oxygen, and then other metals are added to make the kind of steel required. In the electric arc process, scrap iron is usually used. Newly made steel leaves the furnace as a very hot liquid, and is often poured into molds, where it cools and solidifies into blocks called ingots, ready for casting (shaping).

■ Metals

See how many different kinds you can find around the house.

1. Copper *A soft reddish metal that conducts electricity well. Most electrical wires are copper.*
2. Brass *A tough bright yellowish alloy (mixture) of two-thirds copper and one-third zinc.*
3. Stainless steel *A steel that does not rust or stain, containing chromium and a little carbon.*
4. Bronze *A heavy, dull alloy of copper and tin.*
5. Nickel *A magnetic metal often mixed with copper and zinc to make cheap "silver" jewelry.*
6. Gold *A soft yellow metal that never corrodes; used in jewelry, dental work, and electronics.*
7. Iron *A strong, soft, easy-to-use metal that rusts when exposed to air and water.*
8. Chromium *A hard silvery metal often coated electrically onto other metals as protection.*
9. Steel *A very tough alloy of mostly iron and carbon, very widely used in cars and ships.*
10. Tin *An easily shaped, noncorroding silvery metal, used with other metals to make cans.*
11. Aluminum *A light whitish metal, used in aircraft and window frames.*
12. Silver *A shiny metal used in jewelry and also on photographic film.*

EXPERIMENT
Measuring density

Because a block of iron weighs more than a block of wood the same size, iron is said to be denser than wood. Every substance has its own density, which is how much a certain amount of it weighs. This experiment shows how to work out the density of different solids.

YOU WILL NEED
● *plastic bottle* ● *measuring cup* ● *craft knife*
● *scales* ● *pen* ● *graph paper* ● *plastic tube*

1 CUT THE TOP off a plastic bottle, make a hole a little way down, and fix the tube in with modeling clay, making the seal as watertight as you can. Always cut away from yourself. Parents of younger children should do the cutting for them.

2 PUT A CONTAINER under the spout, and fill the bottle with water until it pours out of the spout. Now select the object to be measured.

3 PLACE AN EMPTY measuring cup under the spout, and immerse the object fully in the bottle. Note how much water spills into the cup.

4 TAKE THE OBJECT from the water, dry it, and weigh it. To calculate the density, divide the weight by the volume of water in the cup.

■ Density

When scientists talk about how much something weighs, they usually talk about its "mass" rather than its weight. Scientifically, you should use the word "weight" only when you are talking about the force gravity exerts on a particular object. So density is not the weight but the *mass* of 1 cubic centimeter of the substance. Water, for instance, has a density of 1 gram per cubic centimeter. Gases are very light and have low densities — water is almost 1,000 times as dense as air. Most solids, however, are much denser than water. Gold, for example, is almost 20 times as dense as water.

Solids

ROCK IS SOLID, water is liquid, and air is gas — but they do not have to be. Every substance in the world is a solid, a liquid, or a gas, yet, given the right conditions, each can change into any of these "states of matter": solid, liquid, or gas. Even rock melts to liquid lava in the heat of a volcano; water freezes solid on very cold nights.

It all has to do with the tiny particles — atoms and molecules — that make up every substance (p. 26). In a solid, the particles are tightly knit together, like the bricks in a wall. So solids are fairly rigid and tend to keep their shape. You can pick up many solids but not water. When the solid gets hot, however, the bonds between the particles begin to break up and the solid melts to a liquid.

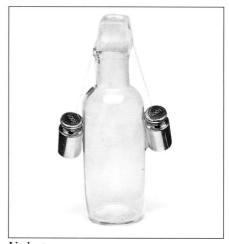

Under pressure
Besides heat, pressure melts solids, as this simple experiment shows. Hang a weighted wire over an ice cube. The wire cuts slowly through the cube as the pressure melts the ice.

EXPERIMENT
Melting point

Adult supervision is required.
The melting point is the temperature at which a solid turns to a liquid.

YOU WILL NEED
● heater (p. 10) ● thermometer (p. 8)
● test tube ● heat-proof beaker or pan
● test solids — e.g., butter, chocolate, Jell-O, wax ● insulated tongs

How to do it
Hold a test tube containing a small quantity of the test solid. Heat the water slowly in the beaker or pan. Keep shaking the tube gently. As soon as the solid begins to melt, check the temperature of the water. Repeat the experiment twice for each solid. Beware: the test tube can get quite hot. Hold it with insulated tongs instead of just your fingers.

EXPERIMENT
Solids are smaller

Adult supervision is required.
When liquids cool, they solidify. Solids are generally denser and their atoms and molecules more densely packed, so they take up less space than a liquid. This experiment shows how butter, like most substances, shrinks when it solidifies, as the molecules pack closer together. Parents of younger children should melt the butter, and do the pouring for them.

You Will Need
- heater (p. 10) ● small, narrow jar ● pan with insulated handle
- butter ● knife

1 MELT SOME BUTTER gently in a pan over the alcohol burner, taking care to heat it only enough to melt it. Stir it until it is all thoroughly melted. Once all the butter has turned to liquid, blow the alcohol burner out.

2 POUR the melted butter into a small, narrow jar. Fill to the brim, and place in the refrigerator to cool.

3 WHEN YOU TAKE the jar out again, you will see a deep hole in the butter, because solid butter takes up less space.

▓ Water is strange
Unlike every other substance, water expands and becomes less dense when it freezes. This is because when water is ice, the bonds between molecules are less tightly knit than when it is liquid. This is just as well. Because ice is less dense than water, it floats to the top. So when rivers and lakes freeze in winter, ice only forms a layer on the top. This layer protects the water beneath and keeps it from freezing. If ice sank, rivers and lakes would freeze solid, killing water plants and animals.

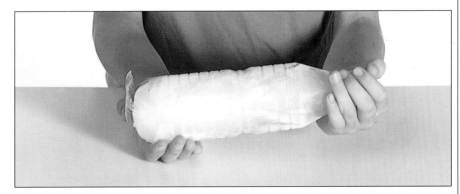

Ice breaker
You can show that water expands on freezing by filling a plastic bottle with water and placing it in a freezer. As the water freezes, it expands with such force that it shatters the bottle completely. This is why water pipes often crack in winter.

Liquids

WHEN A SOLID MELTS and becomes liquid, it changes dramatically. The network of bonds that held its molecules together breaks apart, and the molecules wander all over the place. Indeed, they wander almost anywhere they can, which is why liquids flow into any shape. While molecules in a solid are like a troop of soldiers always moving in step, molecules in a liquid are like dancers on a crowded dance floor. Like dancing couples, clusters of molecules move around in any direction, but not far, for they are constantly bumping into other clusters. When a liquid heats up, the clusters move around faster and faster until eventually some break away to become gas. This is called evaporation.

Water is by far the most common liquid. In fact, at normal temperatures there are few different liquids, apart from water, oils, and spirits — although many other substances dissolve in water (p. 34).

EXPERIMENT
Finding holes in water

Although the molecules of a liquid are in close contact, there is actually considerable space between them — unlike in a solid. You can prove this by dissolving sugar in an apparently full cup; the sugar will fit into the spaces between the molecules. Warm the water to dissolve the sugar more easily.

YOU WILL NEED
● bowl of sugar
● cup of warm water
● teaspoons

1 FILL A CUP or any other container to the very brim with warm water. You need a steady hand to add the last drops so that it fills but does not overflow.

2 VERY CAREFULLY drop a spoonful of sugar into the water, and wait for it to dissolve. Now add another spoonful, taking care not to spill the water.

3 YOU WILL FIND that you can add two or more spoonfuls. When the spaces between the molecules are finally filled, any more sugar makes the cup overflow.

EXPERIMENT
Viscosity

All liquids flow to fill the space in which they are contained, but some liquids flow more easily than others. Oil, for example, flows more slowly than water, and syrup flows more slowly still.

A liquid's resistance to flow — its stickiness — is known as "viscosity." A highly viscous liquid is one that flows only slowly. It is much harder to push things through a very viscous fluid. So you can measure viscosity by how easily things move through it, as we do here. Interestingly, pushing harder only makes the liquid resist more, for pressure increases viscosity. This is because pressure squeezes the "holes" in the liquid and makes it harder for molecules to move. Heating a liquid makes it less viscous, because the holes get bigger.

YOU WILL NEED
● marbles ● jelly jars ● test liquids, such as milk, cooking oil, glycerol, wine, water

EXPERIMENT
Boiling point

*Adult supervision is required:
boiling liquids can be very dangerous.*
Every solid melts at a certain temperature, so every liquid has its own "boiling point." Although solids melt only at one temperature, liquids can change to gas (evaporate) at almost any temperature. Boiling point is not the point at which a liquid turns to gas; it is the highest temperature the liquid can reach. Once a liquid reaches boiling point, it cannot get hotter. The boiling point of water is 212°F (100°C). Like melting point, it goes up with rising pressure.

Measuring boiling point
To find a liquid's boiling point, pour a small quantity of the liquid into a test tube — just enough to cover the bulb of the thermometer. Holding the tube with tongs and pointing it away from your face, heat it over the alcohol burner and watch the temperature rise on the thermometer. Once the liquid boils, the temperature will stay the same. This is the boiling point.

You Will Need
- heater (p. 10)
- test tube ● goggles
- insulated tongs
- thermometer (p. 8)
- test liquids ● goggles

Beware! Boiling liquids are dangerous if handled carelessly. Test only edible liquids, and boil only tiny quantities of oil, since these can reach very high temperatures. Your thermometer must have a range up to 680°F (360°C).

1 To test for viscosity, fill clear jelly jars with equal amounts of the liquids and place them against a white background so that you can see clearly what happens. Put two of the jars together, take a marble in each hand, and hold one over each jar. Then release them at exactly the same moment. You could get a friend to do this while you watch.

2 Watch closely which marble reaches the bottom of the jar first. The marble that takes longer to fall is in the more viscous liquid. Repeat the test, changing one liquid each time, until you can put the liquids in order of viscosity. Remember: both marbles must be released at precisely the same instant if the test is to work properly.

Gases

UNLIKE SOLIDS AND LIQUIDS, gases cannot often be seen or touched, although some, like ammonia, have a strong smell. Yet gases too are made of molecules and atoms. It is just that the molecules of a gas are very far apart, and move about at high speeds in all directions. If gases are cooled enough, the molecules slow down and the gas condenses to become a liquid, just as the steam from a kettle condenses into water droplets on cold surfaces. Most gases, though, condense only at very low temperatures indeed. Oxygen, for instance, condenses at −183°C.

Molecules in motion
Shake beads violently in a plastic jar, and you get some idea of the way molecules shoot about in a gas.

■ Molecules in motion

Even with the most powerful microscope, it is impossible to see the molecules in a gas. Yet we know they are there because of the way gases behave. A good example of this is the way that gases mix very easily. When a bottle of perfume is opened in one corner of a room, its scent can soon be detected throughout the room. The best explanation for this is that both the scent and the air in the room are made of molecules, and the molecules move about rapidly. Scent molecules spread quickly through the air, and you smell the perfume because some reach your nose. This rapid mixing is called "diffusion."

Diffusion occurs in liquids as well as gases, but more slowly; in solids, it takes many years. Gases can also diffuse through apparently solid skin, especially if it has tiny holes. Our bodies rely on diffusion to absorb oxygen from the air into the blood through the lungs. Every balloon eventually goes down because the air inside gradually seeps out through the rubber by diffusion.

EXPERIMENT
Gases and volume

This experiment shows how, because molecules in a gas are farther apart, gases take up more space than equivalent amounts of solids or liquids. Mixing bicarbonate of soda with vinegar causes a chemical reaction that releases the gas carbon dioxide.

YOU WILL NEED
● *vinegar* ● *bicarbonate of soda* ● *balloon*
● *spoon* ● *narrow-necked bottle*

1 TAKE A SMALL narrow-necked bottle, and carefully pour in vinegar until it it is about one-quarter full. Warming the vinegar will speed up the reaction.

2 POUR BICARBONATE of soda into the neck of the balloon through a funnel. Tap the funnel occasionally if it clogs up. Fill the ball of the balloon with the soda.

Diffusion in action
These photographs (left and right) show a famous demonstration of diffusion in gases. Two jars are stood mouth to mouth, separated by a glass plate. One contains air, the other dark brown bromine vapor (left). When the glass plate is removed, the bromine spreads slowly upward — even though bromine is heavier than air. Meanwhile, the air spreads down until the entire space is filled with an even mix of bromine and air. It takes about an hour for this to happen, even though each molecule moves so fast that it can travel up and down the jar 400 times a second. Mixing takes a long time because molecules keep bumping into each other and heading off in various directions.

■ DISCOVERY ■
Robert Boyle

ROBERT BOYLE was an Irish scientist who studied gases in the 17th century. His most famous work involved the "pressure" of a gas — which is actually the combined force of all its moving molecules as they hit anything in their way. Boyle showed that when you squeeze a gas, the pressure increases in proportion, provided the temperature stays the same. This is because more molecules are squeezed into a smaller space.

Blow-up
As soon as the bicarbonate meets the vinegar, it begins to fizz as carbon dioxide gas is released, slowly inflating the balloon.

3 KEEPING THE BALLOON hanging down, stretch the neck over the bottle neck. Once it is secure, lift the balloon quickly so that the bicarbonate falls into the bottle — shake it if necessary.

Inside the atom

EVERY SUBSTANCE IN THE UNIVERSE, from green peas to distant planets, is made up entirely of minute particles called atoms. Atoms are so small that it is possible to photograph them only with extraordinarily high-powered "scanning tunneling" microscopes. The period at the end of this sentence alone could contain well over 2 billion atoms. Yet despite their small size, atoms are mostly empty space, for inside every atom are a variety of even tinier particles, like minute planets in space. These subatomic particles, though, rarely exist on their own; in any substance, the atoms are the smallest particles that can exist by themselves. Break up an atom, and the substance no longer exists — just as, if you smash a brick, you have nothing left to rebuild a wall with.

Smashing the atom
Scientists once believed atoms were the smallest possible particles. Then, in 1919, Ernest Rutherford split one apart with this radium gun.

■ The particle zoo

In 1920, people thought atoms contained just three types of particle: protons and neutrons in the nucleus and electrons outside. Since then, however, scientists have discovered hundreds of others by smashing atoms to bits. These new particles exist only momentarily, but their existence shows that protons and neutrons are not the ultimate building blocks of the universe as scientists once thought. In fact, they now believe there are just two basic kinds of particles: quarks and leptons. Quarks are what protons and neutrons are made of. Leptons are what electrons and similar particles are made of. A third kind, called "gauge bosons," are what hold all these particles together.

Every subatomic particle has an "antiparticle" that is its mirror image, though just as real. This photograph shows the tracks made by an electron (green) and a positron (red), the electron's antiparticle, in an atom-smashing machine.

Atom model

To understand how atoms of different substances are arranged, you can build this model of the atom and its electron shells. Beads representing electrons can be added or taken away to model different elements (pp. 28 and 186). Of course, atoms are not really like this inside; it is simply a convenient way of thinking about them.

YOU WILL NEED
● colored cardboard ● square board ● pencil and compass ● scissors ● craft knife ● beads ● short plastic tube ● adhesive

1 CUT THIN SLICES from the plastic tube, 37 in all. These hold the beads in place on the board. Cut the slices as flat as you can so that they sit properly on the board.

2 DRAW FOUR CIRCLES on the cardboard like those shown in the materials, using a compass and pencil. Cut them out with scissors. These form the different electron shells. Alternate the color for each of the shells.

3 STICK THE CIRCLES onto the board, starting with the largest. It is simplest to stick them on top of each other. But for a neater finish, cut away the inside of each larger circle so that they fit flush.

4 STICK THE TUBE RINGS onto the board along the edge of each in the pattern shown by the beads below. You need 1 in the center, 2 in ring 2, 8 in ring 3, 18 in ring 4, and 8 in the outer ring.

Using the model
Put one large red bead in the center to represent the nucleus of the atom. Blue beads represent electrons. To use the model to represent any element, always fill up the places on the board from the center out — leaving no gaps — until you have the right number of electrons, given by the element's atomic number.

Atom shells
Atoms have at most seven electron shells, and there is a limit to the number of electrons that can fit into each. In the first, there is room only for two; in the second, eight; in the third, eight. After that it becomes very complex. If there are four or more shells, the third can hold up to 18. But the outer shell never has more than eight.

Parts of an atom

In the center of every atom is a tiny but very dense nucleus containing two types of particle: protons and neutrons. Protons have a positive electrical charge; neutrons have none. Each atom of a chemical element has a particular number of protons, known as its "atomic number."

Around the nucleus spin much tinier, negatively charged particles called electrons, like planets orbiting the sun. But electrons are not solid balls like planets; they are bundles of energy, moving as fast as the speed of light. In any atom there are nearly always as many electrons as there are protons, so their electrical charges cancel each other out, making the atom neutral. But it is the electrical attraction between the negative charge of the electron and the positive charge of the proton that holds the atom together (p. 148).

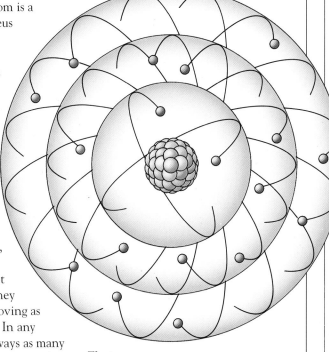

Electrons
Electrons spin around the nucleus in different layers or shells, usually in pairs, with one going clockwise and the other counterclockwise. They can move anywhere in their shell, and sometimes they can even jump from one shell to another. When they jump inward, they give up energy; to jump farther out, they must take in energy.

Elements and compounds

THERE IS SUCH AN INFINITE VARIETY of substances in the world, it is hard to believe they have much in common. Yet everything around you is made from just 92 basic chemicals, called "elements," each with its own kind of atom and unique character. Only the number of protons and electrons distinguishes the atoms, but this has a profound effect on their nature.

Some substances, such as gold, contain just one element. Others contain two or more elements joined together. These are called "compounds." Table salt, for example, is a compound of the elements sodium and chlorine. Compounds have different properties than the elements that make them up. Sodium burns and sputters when put in water; chlorine is a thick green gas!

Molecules

Some atoms can exist alone; most join others to form molecules, the smallest particle of any substance. Compounds are made of identical molecules, each with the same combination of atoms — a salt molecule has one sodium atom and one chlorine atom. Some molecules are made of hundreds of atoms. Shown here are some simpler molecules with colored balls as atoms.

Water
All water molecules consist of two hydrogen atoms bonded to one oxygen atom.

Hydrogen has the lightest atom — just one electron and one proton. With an empty space in its outer shell, it reacts readily with other elements.

Like hydrogen, helium is normally a gas. With just two protons and two neutrons in the nucleus and two electrons, it too is very light.

Elements

Just 92 chemical elements occur naturally. Their characters depend almost entirely upon the makeup of their electron shells. Those that have only one electron in their outer shell, for instance, have many chemical properties in common. These elements, including lithium, sodium, and potassium, tend to combine easily with other elements. Similarly, elements that need just one more electron to complete their outer shells, such as fluorine, chlorine, and bromine, are alike in character. These are called "halogens."

Some elements, such as helium, neon, argon, and krypton, have their outer electron shells completely full. This makes them very stable and slow to react with other elements. They are mostly gases, called the noble gases, and glow brightly when an electric current is passed through them. Neon tubes, for example, glow bright red.

All the elements have been grouped into families like these by being written down in the order of their atomic number — that is, the number of protons in the nucleus (p. 26). This grouping is known as the Periodic Table (p. 186). By looking at an element's position in the Periodic Table, scientists can learn a great deal about how it will behave and also predict how elements yet to be discovered will behave.

Carbon is found in all living things. With four electrons in its outer shell, it has four "vacancies" to form complex compounds with other elements.

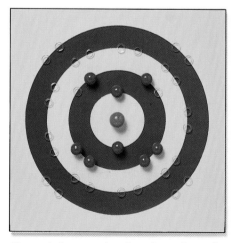

Oxygen is the gas we breathe to live and one of the most common elements. It combines with hydrogen to form water.

Chlorine
In chlorine molecules, two chlorine atoms are bonded "covalently." This means they are joined by sharing electrons in their outer shells.

Carbon dioxide
Molecules of the gas carbon dioxide have one carbon atom joined with two oxygen atoms.

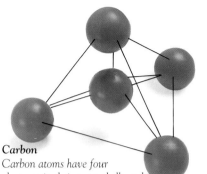

Carbon
Carbon atoms have four electrons in their outer shells and so are four short of the full complement of eight. These "gaps" mean they can bind with other atoms in many different ways. In diamond, which is a form of carbon, each carbon atom joins with four others in a "tetrahedron" shape — a pyramid with three sides and a base.

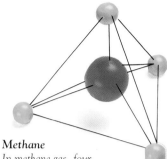

Methane
In methane gas, four hydrogen atoms share their electrons to complete the carbon's outer shell. This is why methane's "chemical formula" is CH_4 — C for carbon, H for hydrogen, and 4 because there are four hydrogen atoms in each molecule. All formulas are written this way to show how many atoms there are in the molecule.

DEMONSTRATION
Iron, sulfur, and iron sulfide

Iron is a shiny grey metal, while sulfur is a bright yellow non-metal. When heated together, the atoms of these two elements react to form a compound called iron sulfide or iron pyrites — also known as "Fool's Gold." Iron sulfide is shiny and yellow like gold, but it's not a metal. Only elements — and not compounds — can be metals.

Test tube

Iron sulfide

▦ How iron sulfide is made
Iron sulfide occurs naturally and can be found in sedimentary rocks. It can also be created in a laboratory. This is done by mixing roughly equal proportions of iron and sulfur particles in a test tube, then heating the mixture to cause a reaction.

ılfur

ron

Iron and sulfur mixture

▦ The scanning tunneling microscope
Atoms are very small, more than a million times smaller than the thickness of a human hair. Even with the most powerful optical microscopes, it is impossible to see atoms. In 1981, the scanning tunneling microscope (STM) was invented. This device, which magnifies objects 10 million times, allows scientists to observe the atoms in molecules. Its probe delicately scans the surface of a molecule, while its tip follows the exact contour of each atom and traces its shape. A scanning tunneling microscope linked to a virtual reality system enables the scientist, below, to see and to feel the atomic details of a molecule.

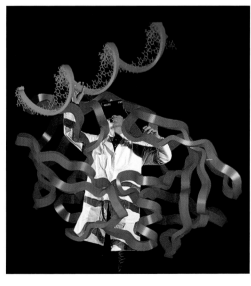

Chemical reactions

WHEN YOU WATCH A CANDLE BURN, a nail go rusty, or a cake rise in the oven, you are seeing chemicals reacting together. Chemical reactions are taking place in the world around us all the time, some natural, some man-made. Sometimes there may be just two elements or compounds involved. Sometimes there may be many more. But whenever a reaction takes place, at least one of the elements or compounds is changed — often irreversibly — and new compounds are formed. Chemicals can meet without

Try writing with lemon juice, which burns before paper. Ask an adult to wave the paper over a hot toaster. The writing will soon become visible.

reacting; a new compound shows that a reaction has occurred. There are many different types of reactions. The experiments here illustrate some of the most common reactions.

Ionic bonds
Chemicals' readiness to react together depends on their atomic structure. The models here show electron shells for chlorine (top) and sodium (bottom, the two elements in common salt. Chlorine has seven electrons in its outer shell, leaving it with a vacancy of one; sodium has just one. When they meet, the chlorine atom draws in the sodium electron to fill its gap. This gives the chlorine a bonus electron, making it negatively charged (p. 27), while the sodium is one short, making it positively charged. These charged atoms are called "ions." Because they have opposite charges, chlorine and sodium ions are drawn toward each other and stick together. This is called "ionic bonding."

DEMONSTRATION
Making sodium chloride

If warmed sodium metal is added to chlorine gas, a violent chemical reaction occurs. During the reaction, electrons transfer from the sodium atoms to the chlorine atoms, forming ions. These ions cling together to form a new compound called sodium chloride, also known as common salt.

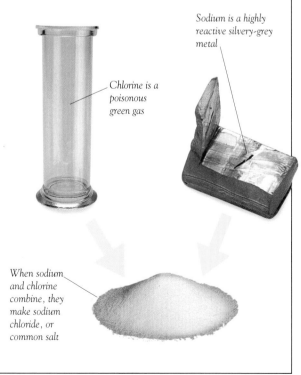

Sodium is a highly reactive silvery-grey metal

Chlorine is a poisonous green gas

When sodium and chlorine combine, they make sodium chloride, or common salt

Identifying elements ■
In every chemical reaction electrons form bonds between the elements that are involved. Each element gives out a different type of light — such as bright green or orange — when its atoms are heated. This fact can be used as a means to identify certain elements in laboratory conditions, using a procedure that is called a flame test. A small sample of an unknown compound is heated in a flame, and the light given off is analyzed to determine which elements make up the compound.

EXPERIMENT
Volcanic reaction

You can have fun with chemical reactions by making this homemade volcanic eruption. It works because bicarbonate of soda reacts dramatically with vinegar to form bubbles of the gas carbon dioxide.

1 ADD RED FOOD COLORING or red ink to vinegar. This will give the "lava" its dramatic, red-hot color.

2 HALF-FILL the plastic bottle with bicarbonate of soda, and stand it up in the middle of the dish.

3 PILE GRAVEL and then sand around the bottle, leaving just the hole uncovered. Smooth the sand into a volcano shape. Then quickly pour all the red vinegar into the bottle, and watch for the eruption!

YOU WILL NEED
- *large dish or tray* ● *funnel*
- *plastic bottle* ● *red food color*
- *vinegar* ● *bicarbonate of soda* ● *sand and gravel*

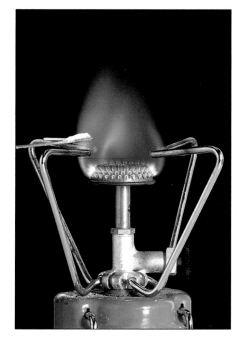

Boric acid
Boric acid burns with a bright green flame because it contains the element boron.

Salt
Salt — the chemical sodium chloride — burns with an orange flame because it contains sodium.

Cream of tartar
Cream of tartar — potassium antimony tartrate — burns with the lilac flame of potassium.

Mixtures

A FEW COMMON SUBSTANCES are made of just one chemical element or compound. Pure water is a compound, for instance. So too is salt. But most are mixtures, including most food, milk, oil, gasoline, and the air. Even tap water is a mixture, for it contains many dissolved substances. The chemicals in a mixture may be thoroughly intermingled, but they do not react together, and, like different marbles in a box, their molecules remain intact. With the right technique, the elements and compounds in a mixture can often be separated out. The experiments here show some ways of separating mixtures.

Distilling Calvados
This spirit is made from cider (which is 95% water and 5% alcohol). To boost the alcohol content, the cider is heated in a still to 70°C, alcohol's boiling point. The alcohol evaporates, condenses at the top of the still, and runs off to make Calvados.

EXPERIMENT
Separating solids

To separate a mixture, you must exploit some property that one substance in the mix has and the others do not. When prospectors pan for gold in sand, they carefully swill the sand in water. Since gold is denser than sand, swilling leaves only gold in the pan. This experiment shows how a mixture of sand and iron filings can be split because iron is magnetic and sand is not.

YOU WILL NEED
● *iron filings* ● *sand* ● *magnet*

■ Metal ores

Metals like iron and aluminum are extracted from the ground as ores, containing many other substances. So to obtain pure iron, for example, one must separate it from its ore by smelting. This involves heating it in a blast furnace until the iron in the ore melts and sinks to the bottom, leaving the unwanted slag on top.

1 POUR THE IRON FILINGS into the sand. Use roughly equal amounts of sand and iron filings.

2 MIX THE SAND and iron together thoroughly. It looks almost impossible to separate them.

3 TO SEPARATE THEM, simply dip the magnet into the mixture, pick up the filings, and brush them into a dish.

EXPERIMENT
Distillation

Distilling is the simplest way of separating two liquids and is one of the oldest chemical processes, widely used for refining oil and making alcoholic spirits. This experiment shows how to separate salt from water.

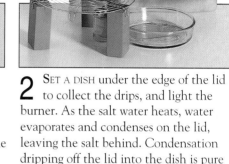

1 MAKE UP a salt solution by dissolving salt in water. Prove to yourself that it contains salt by tasting it. Then set up the alcohol burner (p. 10), and pour the solution into a small saucepan. Cover the pan with a large lid, arranged on a slant as shown in picture 2.

2 SET A DISH under the edge of the lid to collect the drips, and light the burner. As the salt water heats, water evaporates and condenses on the lid, leaving the salt behind. Condensation dripping off the lid into the dish is pure water. Prove it by tasting it.

EXPERIMENT
Chromatography

Chromatography was originally developed to separate plant pigments. But there are now many chromatographic techniques for separating a wide range of substances. The substances are dissolved in water or other solvent (p. 34), and migrate through an absorbent material — paper, chalk, or oil — at different rates.

YOU WILL NEED
● *jars* ● *paper clips* ● *absorbent paper* ● *dropper* ● *inks and food color* ● *dowel rod*

1 MIX TOGETHER two or more of the inks or food colors into separate jars or containers.

2 CLIP STRIPS of absorbent paper to the dowel, and place a small drop of each ink on the end of each strip.

3 SUSPEND THE STRIPS with the ends just in water. The water carries the inks through the paper — some fast, some slow — so they split into colored bands.

Solutions and crystals

IF YOU LOOK AT A GLASS of tap water, you might think it is pure water. It is not. Tap water is a mixture containing traces of various other substances, all of them normally harmless. However, it is a very particular type of mixture called a solution. In a solution, the molecules of a solid, called the "solute," are completely intermingled with the molecules of a liquid, called the "solvent." When you dissolve coffee powder in water, you make a solution, with water the solvent and coffee the solute. Seawater is a solution; so are soup and many other fluids.

When solids dissolve, the solution becomes stronger as more dissolves until it becomes "saturated," and no more will dissolve. If you heat a solution, more will dissolve before it becomes saturated. But if a saturated solution cools down or is left to evaporate, so that it becomes more than saturated, the solute molecules begin to link up and grow into solid crystals.

Heat and solubility
You can prove that solubility — the amount of solute that dissolves — rises with temperature by seeing how much sugar you can dissolve in cold tea, then hot. Hold the lumps just in the tea.

DEMONSTRATION
Growing a crystal

Large crystals can be grown easily from a solution. Here we have grown a crystal from copper sulfate solution, which is POISONOUS. Other types of crystal can be grown from different solutions, such as strong solutions of salt or alum.

Crystallization
As a solution evaporates, it becomes more and more concentrated, until it is saturated and crystals start to grow. A small crystal may be needed to begin the process — otherwise it becomes "super-saturated" before crystallization begins.

The crystal after three weeks

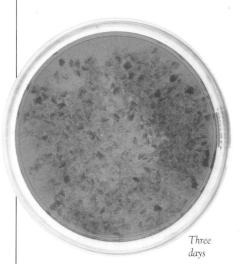

Three days

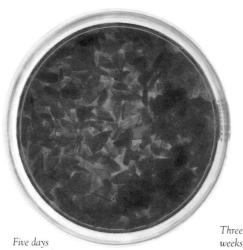

Five days

Three weeks

How a crystal grows
To begin with, a strong solution is made by dissolving a crystalline substance, such as copper sulfate, in water. A little solution is poured

into a shallow dish and left to evaporate. After three days, much of the water has gone, and small crystals have started to grow in the bottom of the dish. After five days, the crystals have

grown in size, and a large one is selected. This is suspended in the original solution and acts as a "seed" for crystallization. After a few weeks, a large, regular-shaped crystal has grown.

EXPERIMENT
Oil and water

Liquids that dissolve in each other are called "miscible;" those that do not are "immiscible," as this experiment shows.

YOU WILL NEED
● *cooking oil* ● *food coloring*
● *water* ● *jug* ● *dropper*

1 POUR A LITTLE WATER into the jug. Then pour in the cooking oil on top. Oil and water are immiscible, so they stay in separate layers. They can be stirred into a mixture of droplets but soon separate again if left to stand.

2 USING a separate dropper or the dropper from the food color bottle, carefully lower one or two drops of food color into the oil. If you are careful, the color will sit in tiny balls because it is immiscible with oil.

3 NOW, USING the end of a spoon, push the balls of food color down through the oil into the water. Once the balls hit the water, they burst in a cloud of color. This is because the food color is soluble or miscible in water.

▮ Crystal structures

Almost every purely solid substance is crystalline, although the crystals may be too small for you to see. Powders, metals, and rocks are all crystals, for example. Ironically, "crystal glass" is not, for glass is actually a special kind of liquid. Crystals seem to come in all shapes and sizes, but this is simply because of the way they grow in clumps. In fact, individual crystals themselves come in seven basic regular shapes. In any substance, each crystal is the same basic shape.

Complex crystals
The green crystals are tourmaline, which contains alum and boron. The pink crystals are lepidolite, which is a type of mica.

Quartz crystals
Quartz is the most abundant of all crystals. Because the crystals are "piezoelectric," a tiny electric current makes them resonate (p. 108) so regularly that they can be used to drive clocks.

Bismuth crystals
Bismuth is one of the few metals whose crystalline structure is large enough to see. The crystals are "tetragonal," which means like two pyramids base-to-base.

Acids and bases

IF YOU PUT A LITTLE LEMON JUICE on your tongue, it tastes sour. So too does vinegar. This is because they are both mild "acids." Lemon juice is citric acid; vinegar is acetic acid. Strong acids, such as sulfuric acid, are highly corrosive; they can burn and sting and even dissolve metals. What all acids have in common is that they contain hydrogen. When mixed in water, the hydrogen atoms lose their single electrons and become positively charged ions (p. 30).

The opposite of an acid is a "base." Weak bases, like baking powder, taste bitter and feel soapy. Strong bases, such as caustic soda, are as dangerous as strong acids. A base that dissolves in water is called an "alkali." Alkalis contain negatively charged ions, typically ions of hydrogen and oxygen, called hydroxide ions. When you add an alkali to an acid, it cancels out the acidity, neutralizing it. The acid and alkali react together to form water and a "salt" (p. 38).

Acid rain
Many industrial fumes contain chemicals called "sulfates," which react with moisture in the air to form sulfuric acid. The resulting acid rain damages trees and plants.

EXPERIMENT
Alkali power

Adult supervision is required: strong alkali must never be touched or swallowed.
Strong alkalis make powerful cleaners. Ovens are cleaned with the alkali potassium hydroxide, for instance. The effect of silver cleaning solution on tarnished silver shows the power of strong alkalis.

YOU WILL NEED
● *item of tarnished silver*
● *bowl* ● *silver cleaning solution*
● *polishing cloth*
● *rubber gloves for safe handling*

1 SILVER TARNISHES because it reacts with hydrogen sulfide in the air. To clean, dip in silver solution, taking care not to get solution on your skin.

2 SOON THE SOLUTION will start to eat away the layer of tarnish, leaving the silver unaffected. Once it is clean, pull it out (don't forget to wear rubber gloves).

Remove traces of solution with a cloth and polish. The silver will be shiny as new.

EXPERIMENT
Acid indicator

Adult supervision is required.
Chemists use "indicators" to test whether a substance is an acid or a base. Indicators work by turning a distinctive color when they are added to an acid or base. You can make your own indicator from red cabbage, as shown, or also from diluted grape juice.

1 TO MAKE YOUR INDICATOR, take a good red cabbage — it must be red cabbage — and chop it up finely on a chopping board. Then boil about a pint of distilled water in a saucepan.

2 ADD THE CHOPPED CABBAGE carefully to the boiling distilled water, and take the saucepan off the heat. Let it stand for half an hour or so until it is completely cool.

YOU WILL NEED
● *red cabbage* ● *distilled water* ● *knife and chopping board* ● *saucepan* ● *sieve* ● *large jar* ● *small jars* ● *substances to test*

3 STRAIN THE LIQUID into a jar, and throw away the used cabbage. The liquid should be a dark reddish purple color. The color will change when you add acids or alkalis.

4 TO TEST IF SOMETHING is acid or alkali, pour a little of your indicator into a small jar. Then add a little of the substance to be tested. A change in color indicates the result as below.

Acid
Acids turn the indicator red — the redder it gets, the stronger the acid. Lemon juice, vinegar, and cream of tartar are all acids. Try orange juice and sour milk.

Neutral
Distilled water dilutes the indicator, but does not change its basic color. This is because pure water is .neutral — neither acid nor alkali. It makes a useful dividing line between the two.

Weak alkali
Tap water is rarely neutral, for it often contains impurities that make it slightly alkaline, turning the indicator blue. Baking soda (sodium hydrogen carbonate) is a salt (p. 38).

Alkali
Alkalis turn the indicator green. Ammonia bathroom cleaner and washing soda (sodium carbonate) are both alkaline. Try milk of magnesia or garden lime.

Lemon

Vinegar

Cream of tartar (tartaric acid)

Distilled water

Tap water

Baking soda

Ammonia bathroom cleaner

Washing soda

Salts and soaps

TABLE SALT, SEA SALT, bath salts, epsom salts — these are just a few of the many compounds called salts by chemists. In fact, most minerals are salts. Typically, they are made when an alkali and acid react together. Metal ions in the alkali take the place of the acid's hydrogen ions, and nothing is left but water and the salt. Two salts can also react to make a new salt.

Not all salts are soluble, but when dissolved, many, calcium salts especially, tend to make water "hard" — that is, water that leaves a chalky feel on your hands when it dries and needs a lot of soap for a good lather. Interestingly, many water "softeners," including soap itself, are also salts.

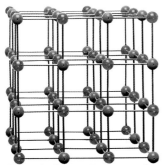

The regular lattice of sodium and chlorine atoms in crystals of ordinary table salt is typical of salts.

▨ Simple salt

Most indigestion tablets contain an acid and a base, such as magnesium carbonate. When they are dissolved in water, the base becomes an alkali and reacts with the acid to form bubbles of carbon dioxide gas (just like vinegar and baking soda, p. 24), and a magnesium salt, dissolved in water. The salt can be recovered by heating the solution until all the water evaporates.

EXPERIMENT
Making bath salts

One easy way to make water softer is to add crystals of washing soda. Washing soda is a salt with the chemical name sodium carbonate. When dissolved in water, it makes the calcium and other salts that make water hard "precipitate" (drop) out of the water. You can use washing soda to make your own bath salts for a soft and foamy bath.

YOU WILL NEED
● *washing soda crystals* ● *mortar and pestle or rolling pin and chopping board* ● *eau de cologne* ● *jug or beaker*

1 CRUSH CRYSTALS of washing soda to a fine powder, either with a mortar and pestle or with a rolling pin on a chopping board.

2 SPOON THE POWDERED washing soda into water, and stir until it is fully dissolved. This is a concentrated solution, so add powder until no more dissolves.

3 ADD EAU DE COLOGNE to make the solution smell nice, and store it in an attractive bottle. Use two large spoonfuls for a soft, foaming bath.

EXPERIMENT
Water hardness

This is a simple experiment to establish how hard the water from your faucets is.

YOU WILL NEED
- tap water ● distilled water ● dropper
- liquid soap ● two screw-top jars ● small jar

1 MIX liquid soap equally with distilled water to make soap solution. Distilled water contains no salts to make it hard.

2 POUR IDENTICAL AMOUNTS of distilled and tap water into each jar. The distilled water is a "control" — that is, something to test your results against.

3 USING THE DROPPER, put one drop of soap solution in the distilled water, screw on the lid, and shake it. Add one drop at a time until you make a foam.

■ How soap works

When soap is made, sodium hydroxide is boiled with vegetable oils. Since the hydroxide is alkaline and the oils acidic, they react to create salts — actually special sodium salts, such as sodium stearate. In the molecules of these salts, the sodium is joined to a long tail of hydrogen and carbon atoms. It is this tail that helps clean your hands. This tail "hates" water, so when you put it in water, crowds of soap tails bury themselves in grease and dirt to escape, lifting the dirt away.

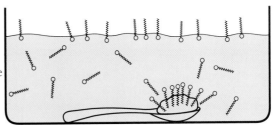

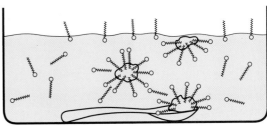

Soap works because the long tails of its molecules are "hydrophobic" — they "hate" water. To get out of the water, the tails bury themselves in grease and dirt, lifting it away.

How hard is it?
Make a note of how many drops of soap solution you needed to make the distilled water foam. Then see how many drops you need to make the tap water foam. The more drops it needs, the harder it is. Try the test with water from other sources.

Organic chemistry

CARBON is the most remarkable of all elements. Its unique atomic structure (p. 29) enables it to link atoms together in long chains or rings in countless ways – – so many that there is a whole field of chemistry, called "organic" chemistry, devoted to studying compounds of carbon. Every living organism, from lichens to people, is made from carbon compounds.

Many carbon compounds are made with hydrogen only, but there are millions of them. These "hydrocarbons" include all the world's fossil fuels — coal, oil, gasoline, gas — fuels formed from the carbon of dead vegetation around 300 million years ago. Other carbon chemicals, such as alcohols and sugars, include oxygen. The largest and most complicated carbon molecules of all are "polymers," from which plastics and many other man-made materials are made.

EXPERIMENT
Making plastic

Adult supervision is required. Most plastics are man-made from oil and natural gas, but you can make your own very simple plastic from casein, the curds in milk, and set it in a mold.

YOU WILL NEED
- *muslin* ● *jars* ● *milk* ● *vinegar* ● *spoon*
- *saucepan* ● *rubber band* ● *heater (p. 10)*

1 WARM 10 FL OZ (.3 LITER) of milk in a pan, but do not boil. Add a tablespoon of vinegar and stir in.

2 A WHITE, RUBBERY MATERIAL, the casein, forms in the milk. Strain it off through muslin held over a container.

3 SQUEEZE THE MILK through the muslin until you have recovered all the solid, rubbery casein.

4 PUSH THE CASEIN into a suitable mold, such as a cookie cutter, or shape it by hand. Then leave it to set.

Polymers and plastics

Some polymers, such as wool and cotton, occur naturally, but plastic and the vast majority of other polymers are man-made. They are actually long chains of smaller molecules called monomers, altered slightly and repeated many times. Polythene, for example, is a chain of 50,000 molecules of a simple hydrocarbon, ethene. These long molecules get tangled up like spaghetti. It is the way they are tangled together that gives the plastic its strength. If the strands are held rigidly together, the result is a stiff plastic like Formica. If the strands can slip over one another, the result is a bendable plastic, like polythene. Forcing the molecules through tiny holes lines them up to form a fiber, such as nylon.

There are now thousands of plastics and polymers, and new ones are being made all the time, for everything from sports clothes to spacecraft. One day, many experts believe, plastics may be made that are strong enough for bridges or clear enough for windows. Even car engines may soon be plastic!

Skiing on plastic

Since they were invented at the end of the 19th century, plastics and man-made polymers have been adapted for more and more uses. In the last few years in particular, sportswear has been revolutionized by materials made from synthetic fibers that combine lightness, strength, and flexibility. Downhill ski racers now race in plastic boots, a plastic crash helmet, and a suit of artificial fibers that is remarkably light, warm, and slippery. Their skis too are made of special carbon fibers such as "kevlar," so strong and light and unaffected by corrosion that they are used in everything from light aircraft to artificial hip joints. In summer, skiers may even practice on artificial, "dry" ski slopes made of plastic fibers.

Plastic brooch

Over a few days, the casein will gradually dry out and harden as the protein molecules bond tightly together. You can speed the process by leaving it on a warm radiator. If you wish to make a brooch, push a safety pin into the back while it is still soft. Once it has set, paint it in your own design.

Refining oil

We obtain many of the most useful organic compounds from crude oil. These "fractions" or parts include gasoline for motor vehicles, diesel, camping gas, and candle wax. Some of the fractions, such as ethene, are used to make plastics (see above). Inside an oil refinery, the different fractions of crude oil are separated from each other so that each can be used. The separation process is called fractional distillation and takes place in a fractionating tower.

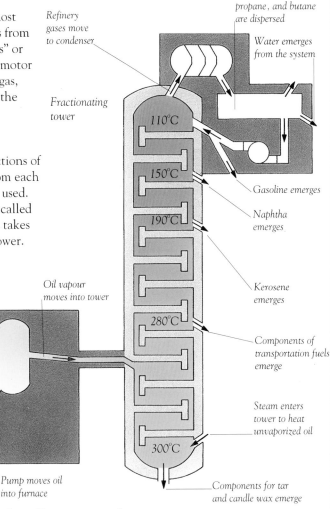

Methane, ethane, propane, and butane are dispersed

Refinery gases move to condenser

Water emerges from the system

Fractionating tower

110°C

Gasoline emerges

150°C

Naphtha emerges

190°C

Kerosene emerges

Furnace heats oil

Oil vapour moves into tower

280°C

Components of transportation fuels emerge

Oil enters system

Steam enters tower to heat unvaporized oil

300°C

Pump moves oil into furnace

Components for tar and candle wax emerge

How fractional distillation works

Crude oil is vaporized in a furnace and enters the fractionating tower at a high temperature. Each fraction becomes a liquid at a different temperature, as it cools at a different level within the tower. The fractions with the lowest boiling points emerge at the top of the tower. As various fractions emerge, they are directed to other parts of the refinery to be processed further.

Energy,
force,
▉AND▉
motion

NOTHING CAN HAPPEN without energy. For anything to move or even change — for butter to melt, for a rocket to fly, for a big dipper to roll, or for us to live — energy is needed. It is energy that creates all forces, and there are forces involved in every movement in the universe.

There are few better demonstrations of Newton's famous laws of motion (p. 62) than a roller coaster, hurtling up and down along its tracks

43

ENERGY

ENERGY TAKES MANY FORMS. Heat energy boils water and keeps us warm. Chemical energy in fuel powers cars, airplanes, and rockets. Electrical energy drives machines, from electric trains to pocket calculators, as well as making lights glow and radios play. But whatever form it takes, it is essentially a capacity for making something happen or "doing work" — whether it is moving something, heating it up, or changing it in some way.

The Sun is the ultimate source of nearly all our energy. Solar cells draw on this energy directly, turning light energy from the Sun into electrical energy — a process that can continue forever, as long as the Sun shines.

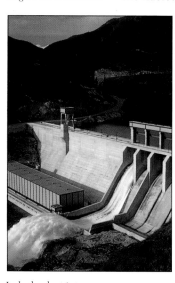

In hydroelectric power stations, running water turns turbines to make electricity to provide an almost inexhaustible source of energy. Unfortunately, building large dams like this can create social and ecological problems.

Scientists talk of two basic kinds of energy, "potential energy" and "kinetic energy."

Potential energy is energy that is not doing anything but is simply stored, ready for action. There is potential energy in wood, coal, and other fuels, for example, and also in food, locked in their chemical makeup. There is potential energy, too, in a wound-up rubber band or a squeezed spring, and energy in the nucleus of an atom (nuclear energy). And anything high above the ground has potential energy because gravity (p. 66) could make it fall at any time.

Kinetic energy is the energy that something has because it is moving — the word "kinetic" comes from the Greek word *kine*, meaning movement. A moving bicycle has kinetic energy. So too has a rolling ball, a speeding bullet, or a falling rock.

■ Converting energy

Almost every form of energy can be converted into other forms in some way. Turn on a light. It gives out light energy. But where has this come from?

The light filament glows because it is heated up by the electrical energy of the electric current. The current comes from the power station. The power station, in turn, produces electrical energy, typically by burning a fuel, such as coal or oil. Burning releases its chemical energy as heat energy, which is used to turn the turbines that generate the electricity.

The conversions can be traced back even further. Coal, for instance, gets its chemical energy from vast forests that grew over 300 million years ago. Coal is actually the fossilized remains of these forests. The forests grew by using light energy from the Sun, by a process called "photosynthesis" which all plants use to convert light into chemical energy for growth.

■ Energy conservation

For every bit of energy, there are similar chains of energy conversions stretching forever backward (and forward). In fact, energy can never be created

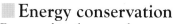

As fossil fuels run low, giant "farms" of wind turbines may be built to exploit wind energy to generate electricity.

Windmills convert wind energy into mechanical energy.

anew. Nor can it be destroyed. It can only be changed from one form to another. This is the Law of Conservation of Energy.

It means that whenever energy is converted from one form to another, the total amount of energy at the end is exactly the same as at the start. Often energy may seem to be lost. In the light bulb, for instance, very little of the electrical energy becomes light. But it is not really lost; it simply becomes heat instead.

In fact, heat is produced as "waste" in nearly all energy conversions. Automobile engines, human and animal bodies, power plants, factories, light bulbs, and millions of other objects produce waste heat. Cars and many other machines usually produce waste sound as well.

■ Energy efficiency

Some machines waste a great deal of energy; others waste very little. The proportion they waste depends on their "energy efficiency." Fluorescent strip lights, for example, are more energy-efficient than light bulbs, because they turn more of the electrical energy into light and lose less as heat. Modern cars and airplanes are shaped "aerodynamically" (p. 128) to ensure that as little energy as possible is lost through friction with the air.

Concern over the potential shortage of energy in the future is encouraging more efficient ways of using energy, such as building Combined Heat and Power (CHP) stations. In ordinary power plants, only about 35 per cent of the fuel is actually converted into electricity; the rest is wasted as heat. In fact, in a typical industrialized country, enough heat is lost from power plants to keep every home in the country warm. CHP stations harness this heat by piping hot water to warm homes and factories.

■ Energy sources

Our daily lives depend on many "energy-saving" devices and machines, from electric tooth-brushes to cars. But these machines only save *us* energy. In fact, they consume a great deal of energy.

The modern world exploits a range of sources of energy, but over 80 per cent comes from burning fuels such as oil, natural gas, and coal. These are all "fossil fuels," formed millions of years ago from the fossilized remains of plants and animals. We are burning up these fuels at an ever-increasing rate — yet they can never be replaced. Estimates vary from year to year, but many experts believe that at present rates of consumption the world's coal supplies will last little more than 200 years and its oil barely 40 years. New reserves are continually being found, but this cannot go on for much longer.

To provide for the future, we need both to become more energy-efficient to eke out reserves longer and also to turn to

A van de Graaff generator uses a rubbing belt (mechanical energy) to create huge sparks of static electricity (p. 148).

more sustainable or renewable alternative energy sources.

Some people believe the answer is nuclear energy. Nuclear energy is the energy that binds together protons and neutrons, the tiny particles in the nucleus of the atom (p. 27). Today, this energy is released by "fission" — which means splitting the nucleus of uranium atoms, the largest atoms of all. But one day scientists hope to use "fusion" — which means fusing together tiny nuclei like those of deuterium (a special form of hydrogen). Unlike fission, fusion should produce a minimum of radioactive waste.

Twenty per cent of the world's electricity is already generated by

Convection currents are one way heat can make things move (p. 54).

Simple conversion of heat to mechanical energy — heat from the candles creates air currents that turn the mobiles.

350-400 nuclear power plants, and some expect this proportion to increase in future. Unfortunately, the problems of disposing of the radioactive waste from nuclear power plants have yet to be solved, and nuclear accidents can be catastrophic in their consequences.

Other, safer alternatives include energy from the sea (wave and tidal power, p. 135), wind energy, solar energy, and energy from the "biomass," which means all living or dead plant and animal matter.

■ Energy and matter

Most of our energy — fossil fuels, solar power, wave power, and so on — comes originally from the Sun. But where does the Sun's energy come from?

Nuclear fusion reactions in the heart of the Sun turn hydrogen into helium and other substances. During this process, matter is lost, and huge quantities of energy are created, mainly as heat and light.

This suggests that matter can be transformed into energy. This is so. Mass and energy are interchangeable — mass is a form of energy and vice versa. Albert Einstein formulated this relationship in a famous equation: $E = mc^2$. E stands for energy, m for mass, and c is the speed of light. Since c is a very large number (and its square even larger), a vast amount of energy must be created from very little matter.

Nuclear power works this way, turning tiny atomic nuclei into huge amounts of heat energy. The nuclear process is the only way energy is created anew — but creating new energy means destroying mass, so the overall mass-energy never changes.

Nuclear power once seemed to offer a cheap, clean, long-term alternative source of energy. But there are problems with disposing of the radioactive waste left by the fission process, and accidents, like that in 1986 at Chernobyl in the former Soviet Union, are always a possibility.

Heat

HEAT IS NOT JUST A WARM FEELING you get from the Sun or a fire; it is molecules moving. The faster molecules in a substance vibrate, the hotter it is. When you put your hand over a heater, the warmth you feel is simply an assault by billions of fast-moving air molecules, spurred on by even faster-moving molecules in the heater itself. The air heats up (and the air heats you) because heat always spreads out. If something is hot, it always passes on its heat to its surroundings, heating them up and cooling down itself. We measure how hot something is by its temperature. But temperature is not the same as heat. Heat is a form of energy — the combined energy of all the moving molecules. Temperature is simply a measure of how fast all these molecules are moving.

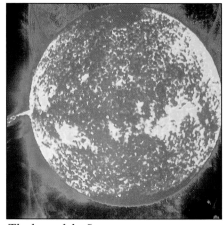

The heat of the Sun
The Sun is our main source of heat. At its core, nuclear fusion reactions boost temperatures to an incredible 14,000,000K. Even at the surface temperatures are over 6,000K (see opposite).

EXPERIMENT
Heating liquids

Adult supervision is required. Heating something raises its temperature. But if you heat different substances equally, they reach different temperatures. Heat equal amounts of water and copper equally, and the copper gets ten times as hot. This is because every substance has its own "specific heat capacity." Try this experiment with liquids such as water and cooking oil — heating the same amount of each liquid for the same time. Which gets hottest? Let the liquids stand an hour before you begin.

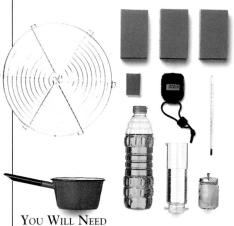

YOU WILL NEED
● *burner (p. 10)* ● *saucepan* ● *measure* ●
thermometer (p. 8) ● *watch or timer*

1 MEASURE OUT a certain amount of cooking oil in a measuring jug or cylinder. Write down the amount.

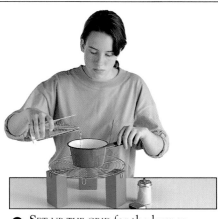

2 SET UP THE GRID for the burner securely on the three blocks. Pour the oil into a clean saucepan and stir.

3 CHECK THE TEMPERATURE of the oil. Light the spirit burner, but do not put it underneath yet.

Warning:
Hot liquids can splatter and cause severe burns. Flammable liquids, such as alcohol, should not be tested in this experiment.

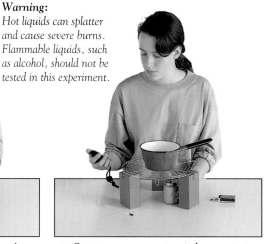

4 SLIDE THE BURNER in and start timing. Heat for five minutes, and check the temperature. Repeat for other liquids.

EXPERIMENT
Heat exchange

Adult supervision is required. Measure the movement of heat between a metal nut and water.

You Will Need
- flask of water ●
- spirit burner (p.10) ●
- thermometer (p. 8) ● nuts
- insulated tongs

1 POUR SOME WATER into the flask, and leave it for an hour. Then check the temperature with a thermometer.

2 CAREFULLY HEAT a metal nut in the burner flame, holding it with insulated tongs. Lower it gently into the flask.

3 WATCH THE THERMOMETER. See how the water temperature slowly rises as heat flows into it from the hot nut. Do more nuts heat the water more? See what effect nuts left in a freezer overnight have on water temperature.

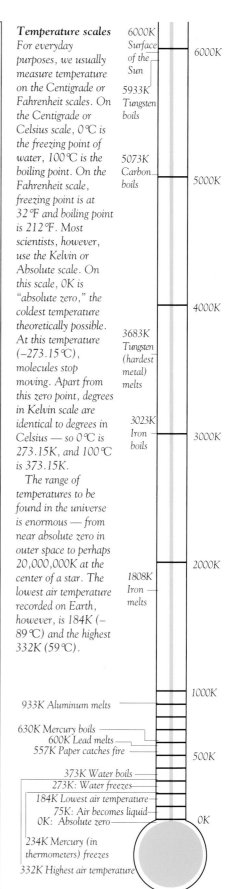

Temperature scales
For everyday purposes, we usually measure temperature on the Centigrade or Fahrenheit scales. On the Centigrade or Celsius scale, 0 °C is the freezing point of water, 100 °C is the boiling point. On the Fahrenheit scale, freezing point is at 32 °F and boiling point is 212 °F. Most scientists, however, use the Kelvin or Absolute scale. On this scale, 0K is "absolute zero," the coldest temperature theoretically possible. At this temperature (−273.15 °C), molecules stop moving. Apart from this zero point, degrees in Kelvin scale are identical to degrees in Celsius — so 0 °C is 273.15K, and 100 °C is 373.15K.

The range of temperatures to be found in the universe is enormous — from near absolute zero in outer space to perhaps 20,000,000K at the center of a star. The lowest air temperature recorded on Earth, however, is 184K (−89 °C) and the highest 332K (59 °C).

6000K Surface of the Sun
5933K Tungsten boils
5073K Carbon boils
3683K Tungsten (hardest metal) melts
3023K Iron boils
1808K Iron melts
933K Aluminum melts
630K Mercury boils
600K Lead melts
557K Paper catches fire
373K Water boils
273K: Water freezes
184K Lowest air temperature
75K: Air becomes liquid
0K: Absolute zero
234K Mercury (in thermometers) freezes
332K Highest air temperature

6000K
5000K
4000K
3000K
2000K
1000K
500K
0K

Heat energy

HEAT IS THE ENERGY of moving molecules and atoms. When you heat something, its molecules move faster. It takes energy to move them faster, but the speeding molecules are themselves energetic. Because heat is energy, it can not only be used for warmth, but changed into other forms of energy. "Fuels" such as coal and oil are burned to release their chemical energy and create heat; the heat can power cars and planes, turn turbines to make electricity, and much more besides.

■ DISCOVERY ■
James Joule

IT WAS the British physicist James Joule (1818–89) who, in the 1840s, proved that heat was a form of energy, and that heat and mechanical energy were interchangeable. He made a machine with paddles rotated in a drum of water by falling weights. As they turned, the paddles warmed the water. Joule measured the temperature change and showed it was precisely related to the weight.

EXPERIMENT
Steamboat

Adult supervision is required.
See how heat energy changes to movement.

YOU WILL NEED
● *tools and sheet balsa as shown* ● *18 in x ¹/₈ in copper pipe* ● *plastic tube* ● *glue* ● *nightlight candle* ● *jar lid* ● *small can or sheet aluminum* ● *paint* ● *pencil or ³/₈ in dowel*

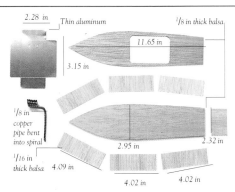

2.28 in — Thin aluminum — ¹/₈ in thick balsa
11.65 in
3.15 in
¹/₈ in copper pipe bent into spiral
2.95 in
2.32 in
¹/₁₆ in thick balsa — 4.09 in
4.02 in — 4.02 in

1 CUT OUT the balsa, and glue the sides to the base. Cut the bows at an angle to mate neatly. Make sure the sides are upright — hold with pins if necessary.

2 NOW GLUE the deck in position. Test for water-tightness in water, and fill leaks with glue. Glue a long strip of balsa down the center of the boat for a keel.

3 USE AN EXISTING shallow can, or cut the aluminum with tin snips, score, and fold into a box shape. Drill two small holes for the copper pipe.

4 CAREFULLY BEND the copper pipe six times around a pen or ³/₈ in dowel. Drill two holes in the boat base aligned with those in the shallow tin.

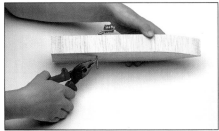

5 SET THE METAL BOX in the boat, bend the pipe at 90°, and slot the ends through the holes. Glue around the pipes. Bend the ends back with pliers.

6 GLUE THE LID inside the metal box, and place the candle on it so that the flame will heat the coil of copper pipe. The metal stops the wood from burning.

▦ Heat engines

Heat engines are of two main kinds. With "external combustion" engines, such as steam engines, the fuel is combusted (burned) outside the engine's moving parts; with "internal combustion," such as jet engines and car gasoline engines, the fuel is burned inside. Illustrated here is a steam turbine, in which a fuel is burned to heat water to create steam. The steam spins fanlike turbine blades attached to a central shaft which turns with the blades. The shaft can be used to turn anything from an electricity generator (p. 166) to a ship's propeller.

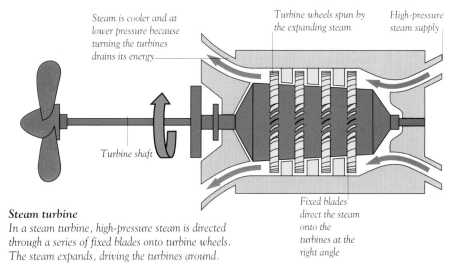

Steam is cooler and at lower pressure because turning the turbines drains its energy

Turbine wheels spun by the expanding steam

High-pressure steam supply

Turbine shaft

Fixed blades direct the steam onto the turbines at the right angle

Steam turbine

In a steam turbine, high-pressure steam is directed through a series of fixed blades onto turbine wheels. The steam expands, driving the turbines around.

Night-light candle

Lid

Heat-tray made from sheet aluminum

Coil bent around ⅛ in dowel

Deck of ⅛ in balsa

Keel made from strip of ¹/₁₆ in balsa

Side view of finished boat

Sides of ¹/₁₆ in balsa

8 LIGHT THE candle with a taper, and before long the boat will start to move steadily across the water, making a gentle put-put sound. What you have made is called a PWE, or pulsating water engine.

How it works

The boat is driven by the candle's heat energy. The flame heats water in the coil, turning it to steam. Steam pushes water from the tubes, driving the boat forward. At once, the steam condenses and rushes back up the tube to be heated again. So the boat sails on in a series of pulses.

7 FIT THE PLASTIC TUBE to one end of the copper pipe, put the boat in water, and suck to fill the pipe with water. Remove the tube, taking care to keep the pipe full.

Expansion and contraction

On hot summer days, overhead cables may sag, doors start to jam, and railroad lines buckle. All these problems are due to expansion. When solids and liquids heat up, they expand because their molecules vibrate more and move farther apart. When they cool down, they contract in the same way. With solids, expansion and contraction are too tiny to see. A bar of steel, for example, gets longer by only 0.0001 per cent for every degree Celsius (p. 47) rise in temperature. But the force of expansion is so powerful that engineers must allow for even these minute changes when building bridges and roofs, and girders can be fixed with red-hot rivets which cool and contract to make a firm joint. Gases expand much more than solids and liquids but less powerfully; if they are confined within a strong, rigid container, their pressure goes up (p. 25) but they do not expand.

Drooping wires
Overhead power and telephone wires must be allowed to sag a little if they are hung from pylons and poles in summer — otherwise they will snap when they contract on cold winter nights. Similarly, railroad engineers leave a gap between lengths of rail to allow for expansion on hot days.

EXPERIMENT
Expansion meter

Adult supervision is required. Some substances, such as wood and plastic, hardly expand at all. Others, especially metals, increase in size a lot. This experiment shows the expansion in a thick wire. Try it with different kinds of wire to see if they expand by different amounts.

You Will Need
- wood for base and blocks
- saw ● cardboard ● straw
- pen ● protractor ● knife
- knitting needle ● night-light candles ● matches ●
- weight and string ● pliers ●
- thumbtacks ● coat hanger wire and/or similar stiff wire

Dial fixed to block with thumbtacks

Cardboard dial with lines drawn using a protractor

Test wire

Drinking straw pointer

Knitting needle

Night-light candles

Weight

Dial

Drinking straw

Test wire

Knitting needle

Wooden block

End-on view

Setting up the meter
Nail two blocks to the base, and tack the card dial to one. Stick a straw on the needle, and lay it over one block. Secure the test wire in a hole in the other block. Hang a weight from the other end.

Using the meter
Set up the meter as shown above, hang the weight on the wire, and light the candles. As the wire heats up and gets longer, it rolls over the needle, twisting the straw pointer farther across the dial.

EXPERIMENT
Hot-air balloon

When air or any other gas gets hot and expands, it gets less dense (lighter) because the same amount of air occupies a larger space. Hot-air balloons fly because they contain warmer, lighter air.

YOU WILL NEED
- *colored tissue paper* ● *glue* ● *brush* ● *scissors* ● *string* ● *paper*
- *ruler* ● *pen* ● *plastic lid* ● *powerful hair drier*

1 MAKE A cardboard template for the balloon panels from the pattern below. Cut out eight panels from tissue paper.

2 GLUE THE PANEL edges together into a balloon shape. This is quite fiddly, so you will need to be patient.

3 CUT A PIECE of paper for the basket. Make firm creases in the paper, so that it will fold easily.

4 FOLD AND GLUE the paper into a box shape. Make a small hole in each top corner of the box with a pencil.

5 THREAD ONE PIECE of very lightweight thread through each of the holes in the corners of the basket.

6 GLUE OR TIE EACH thread to the balloon's mouth, making sure the threads are secure. Allow the glue to dry.

7 HOLD THE TOP of the balloon, and fill it with warm air from a hair drier set on maximum. Once the air inside is really warm, let the balloon go and let it rise into the air.

Helpful hints
Fly your balloon in a large, cold room. If your balloon does not rise, try it without the basket. Attach three paper clips around the bottom, to stop the balloon from toppling over. If it still does not work, try making it bigger.

Panel template
Transfer this shape to a larger piece of paper, using the squares as a drawing guide. Make your balloon as large as you can to help it to fly. (This one is shown small, so that you can see more clearly how it is made.)

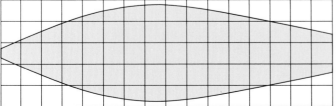
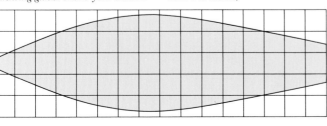

Conduction and insulation

Sᴛᴇᴘ ᴏɴ ᴀ ᴄᴀʀᴘᴇᴛ in bare feet, and it feels warm to the touch. Step on a tile floor, though, and it will probably feel ice cold. The tiles feel colder because they are good "conductors" of heat, and carry body heat quickly away from your feet. Carpet, however, is a good "insulator" and carries heat away only slowly.

Heat conduction is a little like a relay race. When one end of a conductor is heated, its atoms and molecules vibrate faster, jostling neighboring molecules and setting them vibrating faster too. These molecules, in turn, jostle their neighbors, and so heat is gradually passed through the conductor from molecule to molecule, losing a little energy in each handover. Metals tend to be good thermal (heat) conductors, especially copper and aluminum — which is why they are good for saucepans. Air is a good insulator, which is why materials that trap air, including polystyrene and natural fibers such as wool, are usually good insulators too.

Warm birds
Good insulation helps Adélie penguins survive the bitter Antarctic cold. They have an outer layer of oily waterproof feathers, an inner layer of fluffy down feathers, and a thick layer of fatty blubber under the skin. These layers all act as insulators to keep in body heat.

EXPERIMENT
Insulators

Coats keep you warm because they are insulators and prevent the warmth of your body from escaping into the surrounding air. Different materials, though, have different insulating properties. Try this simple experiment to measure the insulating properties of different materials. If you wish, you can make the test more elaborate, by placing the jar in a box and filling the box with different materials. You could try sand, polystyrene beads, feathers, absorbent cotton, newspaper, and so on.

Yᴏᴜ Wɪʟʟ Nᴇᴇᴅ
● *graph paper* ● *pencil* ● *watch* ● *thermometer* ● *jelly jars*
● *warm water* ● *modeling clay* ● *towel* ● *other materials*

Measuring insulation
Make holes in the lids of two jelly jars large enough for a thermometer. Half-fill the jars with warm water. Take the temperatures. Wrap one in a towel. Take the temperatures every two minutes, leaving the thermometer a few seconds for a steady reading. Draw a graph of the results. Repeat the test with different materials around the jar, always starting with the water temperature the same.

Each time you take the thermometer out, seal the hole with a blob of modeling clay

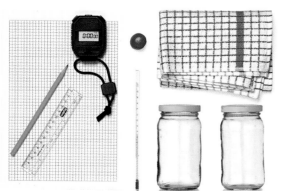

■ Keeping heat in

Few things are better at keeping hot things hot (or cold things cold) than a Thermos flask. It works by cutting down the passage of heat in all possible ways. Inside the casing, there is a flask with a double wall of glass or plastic with a vacuum in between. Since heat can neither be conducted nor convected (p. 54) without molecules, the vacuum is very effective at stopping heat from passing through it. Heat can be radiated through a vacuum, however, so the glass is silvered like a mirror. A thick stopper cuts heat loss in this direction.

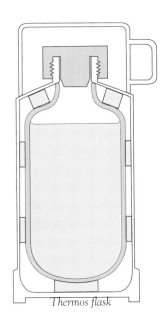

Thermos flask

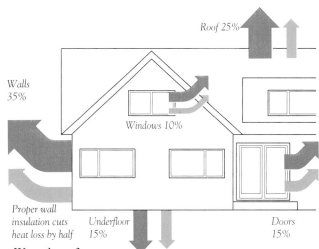

Roof 25%

Walls 35%

Windows 10%

Proper wall insulation cuts heat loss by half

Underfloor 15%

Doors 15%

Warm house?

The need to save energy has focused attention on home insulation. Heat is lost in many ways from the average house, as this diagram shows, but heat loss can be reduced by various insulation methods, including foam-filled cavity walls (which trap bubbles of air between two layers of brick) and roof insulation with fiber and granules.

EXPERIMENT
Conductors

This is a simple way of testing how well different materials conduct heat.

YOU WILL NEED
● *glass beaker* ● *beads* ● *butter* ● *pan of hot water* ● *test items (e.g., wooden spoon, metal spoon, plastic spoon,*

1 STICK A SMALL BEAD to one end of each test item with a small blob of butter, all at the same distance from the far end. Make all the blobs the same size.

2 STAND THE ITEMS upright in the beaker, and pour in about 3 in (7.5 cm) of hot water. Heat from the water is conducted upward through the items. When it reaches the butter, the butter melts and the bead falls off. But since the materials conduct heat at different rates, the beads will fall off at different times. The first bead to fall off shows the best conductor.

Convection and radiation

PUT YOUR HAND above a radiator, and it feels warm. Put your hand by the side, however, and it is much cooler. Some people say this is because heat rises, but this is not strictly true. Heat travels in any direction. It is warm air that rises. What happens is that the radiator warms the air next to it by conduction. As the air warms, it expands and becomes less dense (lighter) than the air around, so it drifts upward like a hot-air balloon. This is called "convection." Cool air is drawn in beneath the rising air, and soon this too warms up and starts to rise. In this way, a stream of rising warm air — a "convection current" — is created by the radiator. Conduction (p. 52) spreads heat relatively slowly; convection currents can soon carry heat throughout a room.

Heat can also spread by radiation. When you stand by an open fire, the warm glow you feel is rays of heat. Heat rays are, like light, part of the electromagnetic spectrum (pp. 78-79), and are called "infrared." Infrared is invisible but, like visible light, can pass through empty space. The Sun heats the Earth by radiation, with heat rays traveling through space at the speed of light (p. 64).

Home central heating
This diagram shows how water circulates around a typical home central heating system. The boiler is likely to be in the basement or on the first floor.

Expansion tank

In a real system, this pipe would join the tank via a "swan-neck" pipe to trap the hot water

Expansion tank overflow

Hot pipes to radiators

Radiator

Cool water down

Hot water up

Boiler

Pump

Return pipes to boiler

■ Using convection

In central heating systems, hot water is circulated by an electric pump. But the pump's task is made easier by convection currents. The boiler is set at the lowest point in the system. The hot water rises through the pipes by convection, as well as pump pressure. As it does, it feeds the radiators, cools, and sinks back through return pipes to the boiler to be reheated. Similarly, in car cooling systems, convection currents help the pump circulate water around the engine. Water warmed by the heat of the engine rises to the radiator, where it cools and sinks through the radiator and returns to the engine.

■ Heat currents

Convection currents occur wherever liquids or gases are heated. For example, the world's wind patterns are created mainly by the heat of the sun warming different parts of the Earth at different rates. This is demonstrated at the seaside, where convection is responsible for sea breezes, as the sun heats the land and the sea unevenly.

Convection currents help heat to spread throughout the gases or liquids present. On a small scale, they occur when vegetables are cooked in a pan of water. The water at the base of the pan is heated by the stove. The hot water rises, losing heat to the water around it, and to the air at the top of the pan. As it looses heat, the water cools and sinks, creating a convection current in the water.

Seeing convection currents
Convection currents can be observed in the beaker below. Small dots of paper rise as the water is heated. The arrows show the rising and falling movement of the water.

EXPERIMENT
Twirling mobiles

Make these mobiles to demonstrate the convection currents created in air above a heat source.

Heat driven
Cut out spirals or fan shapes from discs of foil, and hang them by a thread above a candle flame or radiator. They should spin in the convection currents.

YOU WILL NEED
● candles ●
saucers ● scissors
● wooden base
and dowel ●
thread ●
steel wire
● metal foil
dishes or
aluminum
foil

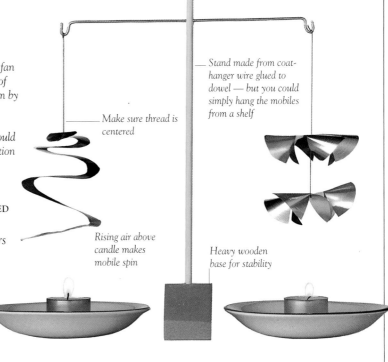

Stand made from coat-hanger wire glued to dowel — but you could simply hang the mobiles from a shelf

Make sure thread is centered

Rising air above candle makes mobile spin

Heavy wooden base for stability

EXPERIMENT
Radiation and absorption

Have you noticed that on a sunny day, you feel cooler in light-colored clothes than in dark ones? This is because, just as light-colored surfaces reflect more light than dark surfaces, so they reflect more of the sun's heat rays away. Dark surfaces, on the other hand, absorb the rays and so get much warmer, as you can see if you try this experiment on a sunny day.

Performing the test
Prepare three jelly jars, leaving one clear, one wrapped in foil, and one painted matte black. Ask an adult to punch a hole in each lid for the thermometer. Put water of the same temperature in each jar, then place them in full sunshine. Take their temperatures every three minutes, blocking the holes with modeling clay after each reading. The black jar should heat up most because dark surfaces absorb the sun's rays well. The foiled jar should heat least, because the shiny surface reflects the heat rays away. But heat losses can alter the result — see caption at right.

YOU WILL NEED
● jelly jars
with lids
● punch
● hammer
● thermometer
● aluminum
foil ● black
paint and
brush
● stop-watch
● tap water ●
modeling clay

Besides absorbing heat quickly, dark surfaces lose heat quickly — so the black jar may lose heat through the shady side almost as fast as it gains it through the sunny side. Glue cardboard to the shady side of each jar to cut heat loss.

FORCE AND MOTION

FORCE AND MOTION are two of the most important concepts in science. Without forces, nothing in the universe would ever happen. Without motion, the universe would cease to exist. They are united in the subject called "dynamics," the science of the way objects move when acted upon by forces.

Of the many achievements of 17th-century science, perhaps none was so important as the new understanding of motion. The philosophers of Ancient Greece had known a great deal about "statics" — that is, about things that are not moving. But when it came to movement, they were largely in the dark.

Galileo was the first to understand that gravity accelerates all falling objects downward to the same degree. Legend says he showed this by dropping two iron balls from the Leaning Tower of Pisa in Italy.

Questions such as why an arrow flew through the air baffled early scholars entirely. They could appreciate the force that fired the arrow from the bow, just as they could understand that a plow moved when an ox pulled it. What they could not understand was how the arrow continued to fly through the air without anything apparently to keep it going.

Galileo

One reason early scholars found it hard to come to grips with motion was their lack of an accurate clock. Without a good clock, changes in position cannot be charted precisely and proper observations about motion are hard to make. Then, one day in 1583, the remarkable Italian scientist Galileo (1564–1642), then a young medical student, was sitting in church when his eye was caught by a bell rope

Riding a bicycle provides a continuous demonstration of Newton's Laws of Motion in action.

swinging back and forth.

Galileo soon began timing pendulums with his pulse and realized that their regular swings gave a simple and reliable way of keeping time. Much later, he designed a pendulum clock, which was built by his son after his death.

Over the next half century, Galileo proceeded to make the observations, experiments, and mathematical analyses of force and motion that laid the basis of the science of dynamics. His ideas were published in his *Discourses Concerning Two New Sciences* in 1638, one of the most important science books ever written. (The other science was the "strength of materials.")

One of his most important insights was into the nature of acceleration. Galileo made the crucial distinction between acceleration and speed — or, rather, velocity. Velocity is sometimes confused with speed, but they are actually different qualities. With speed, the direction of movement is not important. A car's speed is simply

Galileo gained many of his insights into force and acceleration from rolling things down slopes. It is still a good way to do acceleration experiments (p. 65).

how fast it is moving, usually in miles per hour — it does not matter which way it is moving. Scientists call a simple rate like this a "scalar" quantity. Velocity, however, is a "vector" quantity, which means you know the direction as well as speed. This is important because forces always act in a particular direction.

Acceleration

Like velocity, acceleration is a vector; it is simply a change in velocity. After a series of experiments — most notably rolling balls down slopes — Galileo realized that forces control acceleration, not velocity. In other words, a pull or a push makes things accelerate, but does not affect their velocity directly. If this sounds odd, just think of one rocket moving fast through space and another moving slowly. If each is given an identical push, each accelerates by the same amount, but one is still moving faster than the other.

A gyroscope works because, as Newton showed, rotating objects have momentum just like those moving in a straight line.

Galileo also showed that in the absence of a force, velocity stays the same — thus answering why an arrow flies through the air long after leaving the bow. Indeed, it falls back to the ground only because the resistance of the air (a force) slows it down enough for gravity (another force) to pull it downward. This is the principle of "inertia" (p. 58). Galileo went on to show that there is no real difference between something moving at a steady velocity and something not moving at all. Only acceleration makes a difference.

To show how this works, Galileo described a ship at sea. Shut yourself up in a cabin below decks, he suggested, and you will see fish in a fish tank swimming in all directions just as easily when the ship is moving as when it was at anchor. For the fish, the ship's motion is irrelevant. Similarly, when we walk around, we are never aware that the ground beneath our feet is part of a planet whirling around the Sun at 62,137 mph! Nearly 300 years later, Einstein was to develop this idea, called "Galilean relativity," with his special theory of relativity.

How fast?

Other experiments with swinging pendulums led Galileo to a second important insight into force and motion. If something is accelerated, he showed, the acceleration depends on the size of the force and how heavy the object is. A large force accelerates a light object very rapidly; a small force accelerates a heavy object slowly.

Galileo's ideas had made giant steps in the understanding of force

Machines like cranes make work easier — by applying force in one place to overcome another force in another place.

and motion; it was Isaac Newton (1642–1727) who drew these ideas together and laid the basis for the science of dynamics. In his remarkable book *The Mathematical Principles of Natural Philosophy* (published in 1687), he established three important Laws of Motion (p. 62).

■ Newton's Laws

The first two were essentially the same as Galileo's two insights: first, that an object accelerates (or decelerates) only if a force is applied; and, second, that the acceleration depends on the relative size of the force and the object, or, rather, the object's mass (p. 19). In his Third Law, Newton stated that whenever a force pushes or pulls one thing, it must push or pull another equally in the opposite direction — which is why, for example, guns kick back violently when fired (pp. 62-63).

Newton's three laws gave scientists a clear and thorough understanding of the way force and motion were related, and a way of analyzing them mathematically. Indeed, coupled with Newton's insights into the nature of gravity, they seemed to account for every movement in the universe, large or small — from the vibration of the tiniest atom to

Rockets demonstrate Newton's Third Law in action (pp. 62-63).

Spring balances are a simple and accurate way of measuring force.

the movements of planets and stars. Calculations based on Newton enable engineers to build bridges that stand up, and space scientists to guide spacecraft billions of miles across space.

Until the early years of this century, Newton's laws seemed indisputable, and for all everyday purposes they still work perfectly. But in the 1920s, physicists began to delve deeper and deeper inside the atom (p. 15) and learn more and more about the nature of light, electricity, and magnetism (p. 79). As they did, they found Newton's laws no longer working quite so well.

■ Quantum mechanics

On the subatomic level, particles seem to move about and interact according to quite different laws. So to analyze the motion of subatomic particles — such as the way electrons move around the nuclei of atoms — people such as Werner Heisenberg (1901–76), Erwin Schrodinger (1887–1961), and Paul Dirac (1902–84) developed the concept of "quantum mechanics."

Quantum mechanics does not *explain* why subatomic particles move as they do. It simply provides physicists with a mathematical tool for *describing* how they move — and also for predicting how they will move in future. Nowadays, physicists use Newton's laws for most calculations about force and motion, and quantum mechanics when working with subatomic particles.

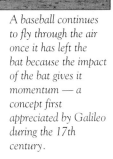

A baseball continues to fly through the air once it has left the bat because the impact of the bat gives it momentum — a concept first appreciated by Galileo during the 17th century.

The explosive force of the gunpowder in a firework accelerates glowing embers in all directions.

Stopping and going

IF YOU PUT THIS BOOK down on a table, it will not move unless someone moves it. This sounds obvious, but it is one of the most important of all physical laws. Every object in the universe has its own "inertia" — that is, it will not move unless something forces it to. So if anything starts to move, you can be sure there is something pushing or pulling it.

Similarly, though not so obviously, any moving object goes on moving forever at the same rate, in the same direction — unless something forces it to slow down or speed up, or pushes or pulls it off course. This is called "momentum." If you have ever been hit hard by a ball, you know what momentum is. In fact, there is no real difference between inertia and momentum, because they both mean there is no change in motion unless a new force is applied.

Just how much force is needed to overcome an object's inertia and get it moving depends on its mass (p. 19). How much force is needed to slow it down or speed it up depends on how fast it is moving as well.

Crash test
There is no better demonstration of momentum in action than a car crashing. Cars are fast and heavy and have enormous momentum — enough to crumple steel in a crash. Passengers have momentum, too, which is why they need seat belts.

■ DISCOVERY ■
Sir Isaac Newton

BORN IN 1642 to a farming family in Woolsthorpe, Lincolnshire, England, Sir Isaac Newton was one of the greatest scientists of all time. His ideas and discoveries in physics, mathematics, and astronomy laid the foundation for modern science.

It was Newton who devised the system of mathematics called calculus, and while at home in the country during the Great Plague of 1665-6, he made many important discoveries about the nature of color and light (p. 94). However, it was his work on gravity and motion that was his greatest achievement. His three laws of motion (p. 62) are still used by scientists to understand the motion of anything from tiny atoms to vast planets. It is said that his ideas on gravity were inspired by seeing an apple fall from a tree, but no one knows how true that is.

Newton's works
Newton's book The Mathematical Principles of Natural Philosophy *(1687) is regarded by many experts as the greatest scientific work ever written. His work aroused much controversy at the time, but he was knighted in 1708. He died in 1727.*

——— EXPERIMENT ———
Overcoming inertia

Spring balances are a very simple way of measuring force. Although designed to measure weight, they can be used to measure force in any direction. You can use the same principle to make a rubber-band-operated force meter to see how much pull is needed to overcome the inertia of a wheeled toy and start it moving, how much is needed to keep it going, and how much to speed it up. You should find that more force is needed to start the toy moving than to keep it going. If you wish, you can calibrate the meter by hanging different weights from it and marking on the meter how far the band stretches.

Spring balance
A simple force meter — the bigger the force, the more the spring stretches.

YOU WILL NEED
● *balsa wood base* ● *rubber band* ●
cardboard ● *screw eyes* ● *beads*
● *thumbtack* ● *string* ● *pen and ruler* ●
card for scale ● *wheeled toy*

EXPERIMENT
Mass and motion

Setting it up
Bind the clothespin's ends with thread. Place two pencils against the peg, aligning them with a ruler. Hold a lighted match on the thread.

Adult supervision is required. You can show how mass, force, and motion are related using a clothespin to fire pencils. Gauge how much force is needed to overcome the pencils' inertia by how far they go. Different-sized pencils travel different distances.

Firing the pencils
When the thread burns through, the spring fires the pencils across the table. Mark how far they go; then repeat with different-sized pencils.

1 ANCHOR THE RUBBER BAND at one end of the base with the thumbtack. Tie one end of the string to it, using a small knot that will slip through the screw eyes.

2 SCREW THE SCREW EYES into the base as below. Thread the string through one eye, the round bead, and the flat bead, and then through the other eye.

3 GLUE THE ROUND BEAD to the thread, and gently pull the rubber band straight. Glue the scale to the base, setting zero where the round bead rests.

4 TIE THE STRING to a heavy wheeled toy on a smooth tabletop. Hold your force meter by the base, and gently pull it until the toy starts to roll. Note the maximum reading shown on the scale by the flat bead, which should brush against the bead. This is the force needed to overcome the toy's inertia.

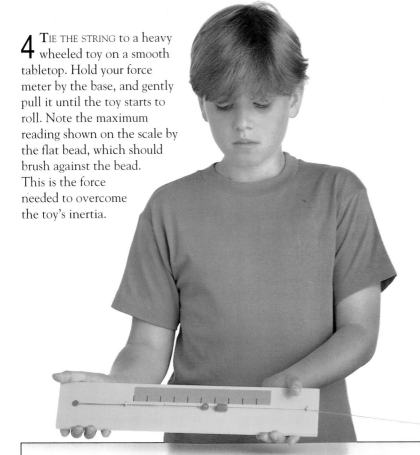

5 NEXT, RESET THE FLAT BEAD to zero; hold the beads in place while you pull on the toy to get it moving. Once it is moving at a steady speed, release the beads and keep pulling the toy at a constant speed. Note the force; it should be much less than was needed to get the toy moving (step 4). Now repeat — only this time speed it up by pulling harder. More force will be needed.

Friction

IF A MOVING OBJECT carries on at the same speed unless some force slows it down, why doesn't it carry on forever? Why, when you throw a ball, doesn't it fly on indefinitely? The answer is that there *are* forces slowing it down. One is gravity, which pulls the ball back to the ground; the other is "friction." Friction is the force between two things rubbing together. When two solids rub together — a metal tray sliding over a table, say — microscopic jagged edges on each surface catch. When a ball flies through the air, its collisions with air molecules cause friction. Friction usually makes things hot, because as the moving object is slowed down, much of the energy of its motion — its kinetic energy (p. 44) — is converted into heat. Rub your hands together briskly, and you can see this for yourself.

Ball bearings
Ball bearings keep friction to a minimum because their round, smooth surfaces roll over each other rather than rubbing — especially when lubricated with oil or grease.

EXPERIMENT
Rocks and rollers

When the Great Pyramids of Egypt were built, some of the huge stones were transported many hundreds of miles to the site. There were no trucks or trains to transport them, and the stones were too heavy to be dragged – the friction between the stones and the ground would have been too great. Instead, the stones were moved on rollers. This experiment shows how effective rollers can be when you want to move something.

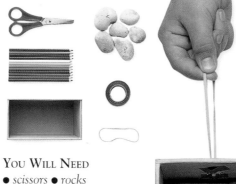

YOU WILL NEED
● *scissors* ● *rocks*
● *round coloring pencils* ● *sticky tape*
● *cardboard box* ● *rubber band*

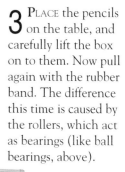

3 PLACE the pencils on the table, and carefully lift the box on to them. Now pull again with the rubber band. The difference this time is caused by the rollers, which act as bearings (like ball bearings, above).

Box moves easily on roller bearings

1 ASK an adult to cut a small hole in one end of the box. Push the rubber band through, and tape it inside the box. Pull the rubber band, and note that the box moves easily.

2 NOW load the box with rocks. Pull the rubber band – does the box move as easily as it did before? The difference is caused by the increased friction between the box and the table.

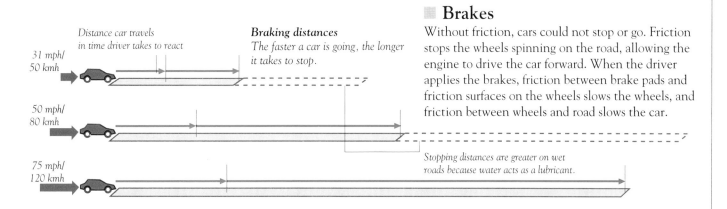

Distance car travels in time driver takes to react

31 mph/ 50 kmh

Braking distances
The faster a car is going, the longer it takes to stop.

50 mph/ 80 kmh

75 mph/ 120 kmh

Stopping distances are greater on wet roads because water acts as a lubricant.

▣ Brakes

Without friction, cars could not stop or go. Friction stops the wheels spinning on the road, allowing the engine to drive the car forward. When the driver applies the brakes, friction between brake pads and friction surfaces on the wheels slows the wheels, and friction between wheels and road slows the car.

EXPERIMENT
Friction slope

Investigate friction created by different surfaces with this adjustable slope.

YOU WILL NEED
- *cardboard scale* ● *pen* ● *protractor* ● *ruler* ● *knife*
- *wooden block* ● *two pieces of balsa or plywood*
- *hinge and screws* ● *screwdriver* ● *thumbtacks*
- *test surfaces*

3 PUT THE BLOCK on the slope. Tilt the slope until the block just starts to slide. Note the angle. Now try the test with different surfaces — felt, shiny plastic, aluminum foil, sandpaper, and so on. At what angle does the block start to slide on these? Then try different lubricants, such as oil and water.

1 SCREW THE TWO PIECES of wood to the hinge to make the adjustable slope. Alternatively, make your own hinge by gluing the wood to a short strip of strong fabric.

2 WITH THE PROTRACTOR draw a scale of angles on the card. Fix this to the base at the hinge with thumbtacks. Set the slope up on a table for the tests.

Action and reaction

WHEN YOU WALK, you may think only your feet do the work. Yet the ground works equally hard. As your foot pushes down, the ground pushes up with *exactly the same force*. If the ground did push any less hard, your foot would sink into the ground as it does into water. If the ground pushed harder, it would shoot you into the air. In fact, whenever anything moves, there is this balance of forces pushing in opposite directions. When a car moves, the wheels push back on the road with the same force as the road pushes forward. When you push your arms and legs on water to swim, the water reacts by pushing back equally hard on your arms and legs. Indeed, for every action there is an equal and opposite reaction. This is Newton's Third Law of Motion.

■ Laws of motion

In 1665, Isaac Newton devised three laws about the way things move that still hold true today in most respects. His First Law is about inertia and momentum (p. 58). It says, essentially, that an object will start to move only if force is applied to it; once moving, it will carry on at the same speed in the same direction unless a force is applied. His Second Law says that the amount an object slows down or speeds up varies with the size of the force and mass (p. 19) of the object. His Third Law is about action and reaction, as described above.

Pushing together

You can see the truth of Newton's Third Law by trying this simple test. You need a smooth floor, roller skates, and a friend on skates (or two friends if you cannot skate). Stand facing each other with hands together as shown. Push your friend away very gently, staying as upright as you can. If you are both equally heavy, you will find, with practice, that instead of pushing your friend away, you both roll back equally. If you are heavier, your friend rolls back farther; if you are lighter, you roll back more.

■ Rocket power

When a rocket roars up into the atmosphere, it is propelled by the blast of hot gases from its tail pushing on the air. But what happens in space, where there is no air to push on? The rocket is still propelled by action and reaction — only there the reaction is between the rocket and the gases rushing from its engine, thrusting the rocket forward and the gases back.

Simple rocket
Rockets like these are used to launch unmanned spacecraft

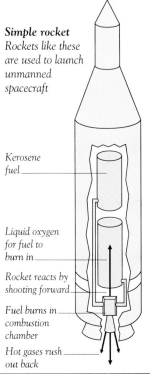

Kerosene fuel

Liquid oxygen for fuel to burn in

Rocket reacts by shooting forward

Fuel burns in combustion chamber

Hot gases rush out back

EXPERIMENT
Water rocket

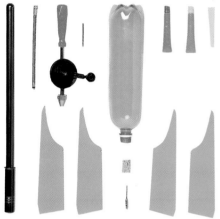

Adult supervision is required.
A rocket is propelled by the action and reaction between its body and burning fuel. This water rocket works on the same basis; it is propelled by the reaction between the water and air pumped into the bottle. It might work without water, but the water provides the mass to give the rocket a huge thrust.

YOU WILL NEED
● *bicycle pump and connector* ● *needle adaptor for inflating basketballs, footballs, etc.* ● *drill and bit* ● *plastic soft-drink bottle* ● *strong glue, such as epoxy (not polystyrene, since this melts plastic)* ● *balsa wood cut into fin shapes* ● *cork*

■ Collisions
A pool table provides a good display of Newton's laws of motion. A moving pool ball has a certain momentum (p. 58), which depends on its mass and the acceleration given to it by the pool cue. When it hits a stationary ball, the law of action and reaction means both balls push each other with equal force in opposite directions. The effect is that the moving ball loses some momentum and the stationary ball gains exactly the same amount. Overall, though, there is no change in momentum. In other words, the momentum after the collision is exactly the same as the momentum before. This is true of all collisions and is called the principle of "conservation of momentum."

Warning: Never put your face or other body parts over the rocket at any time during this experiment.

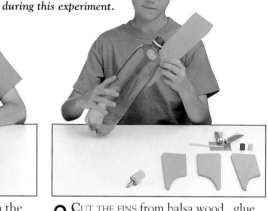

1 ASK AN ADULT to drill a hole in the cork, wide enough for a needle inflator to fit tightly. Push in the needle from the wide end. Watch your fingers.

2 CUT THE FINS from balsa wood, glue them to the rocket, and allow to dry. The fins hold up the base while you pump and help the rocket fly straight.

3 QUARTER-FILL the bottle with water, push the cork in firmly, and connect the pump. Take the rocket to an open space such as a playing field, well away from buildings and overhead wires. Stand the bottle upside down. Keeping your distance, pump air in. The pressure will build up inside until the cork pops out, then ... BLAST OFF!

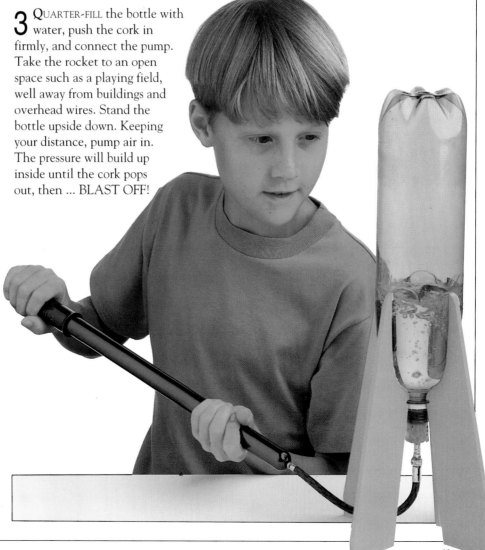

Acceleration

VERY FEW THINGS move at the same speed for long. Moving cars, airplanes, bicycles, and people all eventually speed up or slow down. In fact, nearly everything in the universe is constantly accelerating or decelerating — but by just how much depends on two things: first, the size, or rather, mass of the object changing speed and, second, the size of the forces pushing or pulling it. The larger the force and the lighter the object, the greater the acceleration. This is Newton's Second Law of Motion (p. 62).

How fast?
The speeds we encounter in everyday life, from a snail's pace to a supersonic jet, cover only a small part of the "speed spectrum," ranging up to the speed of light, which scientists believe it is impossible to exceed.

Snail
0.014 ms
0.05 kmh
0.03 mph

Walking pace
1.4 ms
5 kmh
3 mph

Runner
12 ms
40 kmh
24 mph

Cyclist
17 ms
60 kmh
35 mph

Racehors
20 ms
70 kmh
40 mph

EXPERIMENT
Speed trials

With the help of a friend on a bicycle, try these tests on a safe track to show the difference between speed and acceleration.

YOU WILL NEED
● *stop-watch or watch with seconds* ●
markers at 5-meter intervals

1 Get the rider to accelerate slowly from a standing start, and measure the time he passes each marker pole.

2 Take the times at each pole for rapid acceleration.

3 Take the times for a flying start, with the rider staying at a steady speed past all the poles.

4 Compare the times, and work out speed and acceleration.

What is acceleration?
Acceleration is how fast speed is gained. Say you are riding a bike at 2 meters per second (2 ms). You begin to ride faster. After a second, you are riding at 3 ms. After another second, you are at 4 ms. You are gaining 1 ms each second, so your acceleration is 1 ms per second, or 1 ms/s.

0 m   5 m   10 m   15 m   20 m

Standing start
Flying start

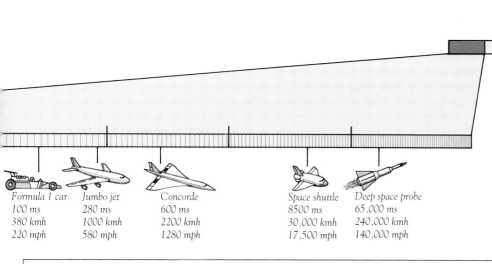

Speed of light: 299,792,458 ms;
1.079 billion kmh; 670 million mph

Formula 1 car	*Jumbo jet*	*Concorde*		*Space shuttle*	*Deep space probe*
100 ms	*280 ms*	*600 ms*		*8500 ms*	*65,000 ms*
380 kmh	*1000 kmh*	*2200 kmh*		*30,000 kmh*	*240,000 kmh*
220 mph	*580 mph*	*1280 mph*		*17,500 mph*	*140,000 mph*

■ Speed and velocity

When scientists talk of how fast something is going, they may talk about "velocity," not speed. By velocity, they mean speed in a particular direction. The figures here are speeds because they are not in any direction. If we referred to a rocket heading for the moon at 50,000 kmh, this would be a velocity.

EXPERIMENT
Acceleration truck

Acceleration means a greater distance is covered in a shorter time. You can demonstrate this with an adjustable slope (p. 61) and a truck with a drip-timer.

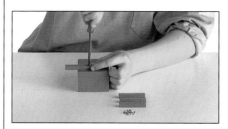

1 YOU CAN STICK the drip-timer onto a ready-made wheeled toy or make a truck as shown here. First screw strips of wood to the block as legs, and cut dowel axles to fit between.

2 THREAD THE SPOOL and washers onto the dowel axles, and fix in place between the legs. Make sure it runs straight and smooth. Cut the top off a detergent bottle for the drip-timer, and glue it to the truck, upside-down.

3 MAKE A PINPRICK in the bottle top. Drip a few drops of ink into glycerine, and pour into the hopper. Adjust the hole to drip about 5 drops a second, then seal the hole with clay.

4 LAY A STRIP OF PAPER on the adjustable slope. Set the truck on the slope. Remove the clay seal. Lift the slope a little way, and let the truck go. When the truck reaches the bottom, seal the drip-timer, lay a new strip of paper on the slope, and repeat, with the slope at a different angle.

YOU WILL NEED
- *adjustable slope (p. 61)*
  For drip-timer:
  - *top cut off detergent bottle by an adult* • *glycerine* • *ink* •
  - *dropper* • *modeling clay to adjust drip hole* • *pin to make hole*
    For truck:
    - *wooden block* • *strips of wood* • *thin dowel* •
      screws • *drill and bit*
    - *thread spools* •
      screwdriver

Distance strips
Increasing distance between drips shows that the truck is moving faster. The gain in distance shows acceleration.

Gravity

WHEN YOU DROP something, it falls faster and faster toward the ground. This acceleration is due to a force called gravity. It is this force that holds us on the ground, and holds the Moon in orbit around Earth. No one knows quite how gravity works, but it is a force of attraction that pulls matter together. Every bit of matter in the universe, from an atom to a star, exerts its own gravitational pull on other matter — but the strength of the pull depends on mass. Big objects exert more pull than small objects — we feel the pull of the Earth's gravity, but not that of, say, a brick.

Astronauts in an orbiting space shuttle float apparently weightless because they and their craft are free-falling around the world together.

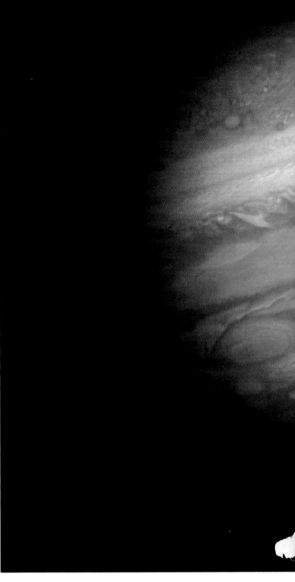

Free-falling skydivers can reduce their terminal velocity by extending their arms and legs out to increase air resistance.

▒ Free fall

The pull of the Earth's gravity — the g force — accelerates everything downward at the same rate, 32.14 feet per second, called the "acceleration of free fall." As an object falls faster, friction from the air increases rapidly. Eventually the air resistance becomes so great that it balances the pull of gravity, and the object can fall no faster. It continues to fall steadily at the same speed, called the "terminal velocity."

■ DISCOVERY ■
Galileo and gravity

The Italian Galileo Galilei (1564–1642) was perhaps the first modern scientist. He invented the telescope, saw the moons of Jupiter for the first time, and made many fundamental discoveries in astronomy and physics. He devised experiments to show that gravity always produces a steady acceleration, and that gravity makes all freely falling bodies drop at the same constant acceleration — no matter how much they weigh. But his ideas brought him into bitter conflict with the Catholic Church, and he was put under house arrest and forced to deny all his work.

*Falling together
Galileo is said to have dropped balls of different weights from Pisa's Leaning Tower to show they fall at the same speed.*

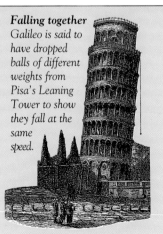

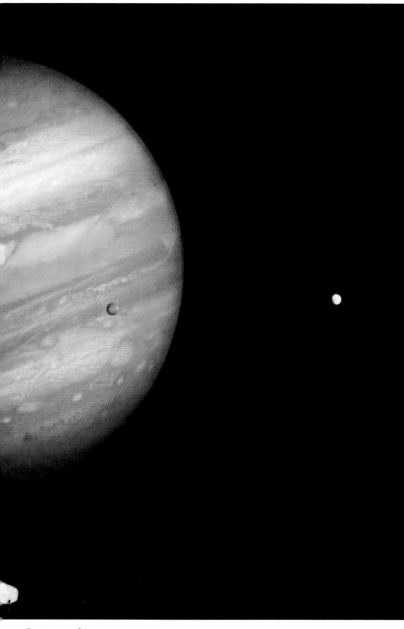

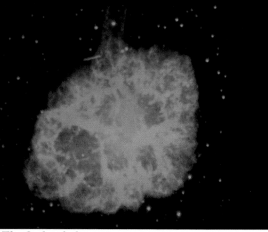

The Crab nebula
The swirling gases of the Crab nebula are lit up by a pulsar. Pulsars are neutron stars so dense that they spin at enormous speed — over 30 times a second — sending out flashes or pulses of radiation like the beam of a lighthouse as they rotate.

Gravity in the universe

Gravity is actually a very weak force — even though it may not seem so when you climb a long flight of stairs. It is not even strong enough to pull together two lumps of lead placed right next to each other. Yet in the physics of space, gravitational forces can be felt over enormous distances and sometimes become so huge that strange things happen. A star, for instance, spends millions of years consuming hydrogen and later helium, emitting light, heat, and other forms of radiation. As these fuels run out, the star collapses under the force of its own gravity. A smaller star may shrink to a cold "white dwarf." A larger star may explode as a "supernova," leaving behind an incredibly dense core of neutron particles (p. 26), called a "neutron star." In theory, a star could shrink so far and become so dense that its gravitational pull is strong enough to suck even light into its grip. This is a black hole.

Jupiter and its moons
The vast planet Jupiter is 318 times heavier than Earth, and its strong gravitational field has captured four large orbiting moons, as well as 10 smaller moons, rings, and dust clouds. Visible here are the moons Io and Europa.

The solar system
The immense gravitational force of our nearest star, the Sun, holds together the nine planets of our Solar System — of which the Earth is just one. The planets hurtle through space at speeds that just balance the Sun's gravitational pull (p. 70), so they are locked into a perpetual circle around the Sun. So massive is the Sun — weighing almost 4.4 million billion billion billion pounds that — its pull is enough to hold Pluto in orbit, 3.7 billion miles away. Moons orbit planets, and satellites and spacecraft orbit the Earth, in the same way. Satellites are not defying gravity in circling endlessly high above the Earth. It is just that they are moving so fast around the world that gravity never brings them any closer. It is a little like throwing a ball so hard horizontally that it never lands.

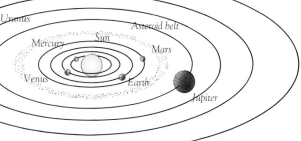

The minute ninth planet, Pluto, orbits so far out that it is not shown here

Turning forces

EVERY FORCE ACTS in a particular direction. Gravity always pulls things down, baseballs go the way you hit them, and so on. Indeed, forces can always be represented in diagrams by straight arrows. So why, then, do so many things in the world turn? The answer is "turning effects." If a force is applied to an object in one place but it is fixed in another, called the "fulcrum," the force turns it around the fulcrum. Whenever you open a door or tighten a nut with a wrench, you are exploiting a turning effect. With the door, the fulcrum is the hinge; with the wrench, it is the bolt on which the nut turns. The size of a turning force is called a "moment," and it depends on how far away from the fulcrum the force is applied, as the experiments below show.

Finding the center of gravity
Here we show how to find the center of gravity (G) for a cardboard car, but you can use the method for any flat object. Let the card swing loosely from a pivot, and hang a length of weighted thread from the same pivot. Mark the line of the cotton across the card. Repeat with two other pivot points. G is where the lines cross.

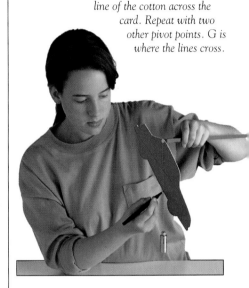

Balanced structure
Turning forces will counterbalance in this sports stand, with equal weight either side of the vertical support so that the structure presses down but does not twist. This is called a cantilever design.

EXPERIMENT
Levers

Levers are simple devices that give you a "mechanical advantage," multiplying your efforts by exploiting moments. The farther away from the fulcrum a force is applied, the greater its effect. Levers are used in countless ways, from scissors to cranes, but they are all of the three basic types demonstrated here.

YOU WILL NEED
● *wood cut as above and in photo* ● *dowel* ● *saw* ● *selection of kitchen weights* ● *string* ● *screws* ● *screwdriver* ● *drill and bits* ● *wood glue* ● *spring balance*

Drill holes large enough for the dowel pegs at regular intervals. Do not glue the pegs.

Glue upright and triangular supports to base, and screw through the baseboard for extra strength

Fulcrum

Load

Warning: Adults should do the cutting and drilling.

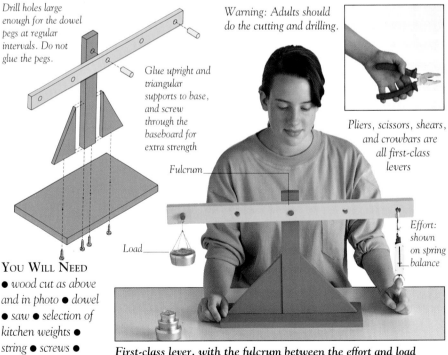

Pliers, scissors, shears, and crowbars are all first-class levers

Effort: shown on spring balance

First-class lever, with the fulcrum between the effort and load
When load and effort are equidistant from the fulcrum, there is no mechanical advantage, and the balance shows the load's normal weight. Applying the effort farther from the fulcrum, or the load nearer, creates an advantage.

EXPERIMENT
Center of gravity

Because gravity pulls on all atoms equally, it pulls any object down as if all its weight were concentrated in one point, called the "center of gravity." When the center of gravity is outside the base of the object, the base can become a fulcrum and gravity a turning force, toppling the object — which is why a bottle will fall right over once tipped far enough. Only if the center of gravity is directly above the base will the object be in "equilibrium" (balanced). This experiment shows how to find the center of gravity.

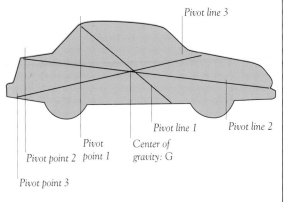

Pivot line 3

Pivot line 1

Pivot line 2

Center of gravity: G

Pivot point 1

Pivot point 2

Pivot point 3

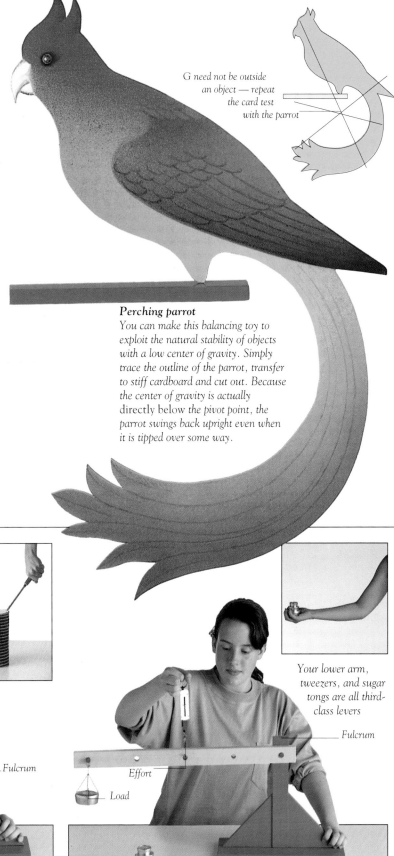

G need not be outside an object — repeat the card test with the parrot

Perching parrot
You can make this balancing toy to exploit the natural stability of objects with a low center of gravity. Simply trace the outline of the parrot, transfer to stiff cardboard and cut out. Because the center of gravity is actually directly below the pivot point, the parrot swings back upright even when it is tipped over some way.

Screwdrivers, nutcrackers, and wheelbarrows are all second-class levers

Effort

Load

Fulcrum

Second-class lever, with the load between the effort and fulcrum
Although load, fulcrum, and effort are all equal distances apart here, the effort is twice as far from the fulcrum as the load, giving it a mechanical advantage. The spring balance shows the reduced effort.

Your lower arm, tweezers, and sugar tongs are all third-class levers

Fulcrum

Effort

Load

Third-class lever, with the effort between the load and fulcrum
Although again load, effort, and fulcrum are all equidistant, this time the effort is half as far from the fulcrum as the load, giving it a mechanical disadvantage. This is useful for tools requiring a delicate grip.

Moving in circles

THE WORLD IS FULL OF THINGS that move in circles – spinning tops, whirling wheels, fans, discs, propellers. Yet forces act only in a straight line. How, then, do these things rotate? The answer is that every spinning object is constantly changing direction. The turning force pushes or pulls it in a straight line. But there is also another force, continuously pulling it in a new direction, inward toward the circle's center. This force is called the "centripetal force" — which may be just the attachment to the center of the circle, like the spokes in a bicycle wheel, or may be an invisible force like gravity. If the centripetal force stops, there is nothing to pull the object around in a new direction, and it will at once fly off in a straight line. This is why you are thrown to one side when your car speeds around a bend.

Feel the force
Every spinning object has angular momentum and resists a change in direction (see below left). Like a gyroscope, it will precess (twist around) when swiveled. You can feel how strong this tendency is by holding a bicycle wheel very firmly at its axle, and getting a friend to spin it fast. Slowly tilt the axle, and the spinning twists the wheel to one side. **SEE WARNING BELOW.**

EXPERIMENT
Gyroscopes

Just as an object moving in a straight line has momentum and resists a change in direction (p. 58), so a spinning object has its own momentum, called "angular momentum." It too resists being moved or stopped.

Gyroscopes are devices a little like tops which exploit this effect. They have a heavy spinning wheel mounted in a pivoting frame. The gyro's spinning wheel tries to stay in the same position as you move the frame. If it is spinning fast enough, it has enough momentum even to resist gravity and refuses to topple when pushed over. Gyroscopes are so stable that airplanes, ships, and even satellites use electrically driven versions to indicate their position for automatic adjustments to the controls. Similarly, the spinning wheels of a bicycle act like a gyroscope and make balancing much easier at speed.

If you forcibly swivel a gyroscope, it does not go the way you push it: it turns at right angles to the way you push. This is called "precession." It happens because your push changes the direction of the gyroscope's angular momentum.

YOU WILL NEED
● *string* ● *toy gyroscope*

BEWARE!
Spinning objects can be very dangerous. Try this only with adult supervision. When spinning the wheel, take great care to hold the end of the axles firmly. Take care not to catch fingers in the spinning spokes.

1 TRY THESE TESTS with a toy gyroscope. Wind the string around the axle, then tug it sharply to unwind it and set the wheel spinning.

2 SET THE GYROSCOPE on its stand, then try pushing it gently in various directions. You will see how well it maintains its stability.

EXPERIMENT
Centripetal force

A cork whirling on a string is kept moving in a circle by centripetal force, shown by tension in the string. Try this experiment to show how this force varies with the speed of the cork to keep it in balance.

YOU WILL NEED
- *drill and bit*
- *string ● cork*
- *thread spool*
- *wooden block or similar weight*

1 ASK AN ADULT to drill a hole in the cork. Tie a loop at one end of the string. Thread the other end through the cork, and then through the spool.

2 TIE THE STRING'S free end to the weight, about 15-20 ins (40-50 cms) from the cork. Make sure the string runs freely through the spool.

■ Leaning into a bend

When a bike corners, its momentum tries to keep it going straight. Since centripetal force (c.f.) acts only through the friction between tires and road, this tends to fling the bike over. The faster the bike corners, the more the rider must lean over to convert more of the bike's weight into c.f.

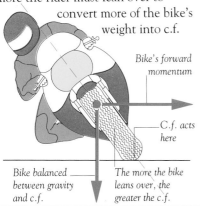

Bike's forward momentum

C.f. acts here

Bike balanced between gravity and c.f.

The more the bike leans over, the greater the c.f.

3 HOLD THE SPOOL. With the weight hanging down, twirl the cork around slowly at first, then faster and faster. As you spin the cork faster, you are increasing the centripetal force, which is causing the circular motion. You can see this increase in force by the way the weight rises.

Gears and pulleys

WHEN YOU'RE CYCLING UPHILL, it helps to have gears. Like levers, gears give a "mechanical advantage" and make pedaling easier. They can give no more power; they simply spread the effort over a greater distance, making the pedals turn faster for each turn of the wheels. Gears can also do the reverse, compressing the effort into a small distance and moving the load farther. Indeed, a bicycle's wheels do just this, because they are much larger than the pedal wheel. This makes pedaling harder, but the wheels carry you farther and faster. Scientists call gears, and any system that changes the balance between effort and load, "machines." Pulleys are also machines.

EXPERIMENT
Gear ratios

Gears are wheels that work together. If the wheels are different sizes, one turns faster, as this experiment shows. The number of times the driving wheel turns for each turn of the wheel being driven is called the "gear ratio." A ratio of 3:1 means the driven wheel turns only once for every three turns of the driving wheel.

YOU WILL NEED
- *jar lids of various sizes* ● *thread spools*
- *glue* ● *thick cardboard* ● *two 1¹/₂-in nails*
- *sandpaper cut into strips to stick around the lid edges and make them nonslippery*

1 TO MAKE AXLES for each lid, push two nails through the card; move one of them each time you change the lids.

2 GLUE thread spools under different-sized lids, and slot over the nails. Adjust the nails so the lids touch.

Changing gear
Try this experiment with a range of different-sized pairs of lids.

3 GLUE A SPOOL to one lid for a handle. Turn once so the other lid turns too. How far does each lid turn?

EXPERIMENT
Model crane

A large and complex machine such as a crane is made up of many simpler machines, including gears and pulleys. The gears and pulleys are powered by an engine or an electric motor. Most modern machines also include complicated mechanisms, such as those that control the movement of an elevator.

This model crane includes both pulley systems, for lifting the jib and the load, and a low-ratio gear for turning the crane. To make it, cut out the shapes as shown on the opposite page, then follow the steps. Measurements are given in millimeters (mm) for accuracy, so use a metric ruler.

1 MAKE THE BOX for the base using pieces A, B, C, and D. Drill holes for the dowels in middle of A on top, and the side of B. The dowels will act as axles for the turntable gear and winder. Cut out the turntable and winder using the pattern in yellow opposite.

2 GLUE together Q, R, and F. Place M through middle holes of Q and R. Attach N and O to one side, and tack the other. Construct the winder jib pieces H, J, and L and, glueing them to L, place this through the spool and bottom holes of Q and R. Tack in place.

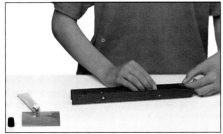

3 DRILL holes through S and T for the swivel. With glue, join S and T to create the jib, using V, W, Y, and Z. Placing 3 through a spool, fix it inside at the front end of the jib using thumbtacks. Glue U inside at the rear of the jib.

4 PLACE X inside the spool, and insert them through the holes in the jib and the top holes in the cradle. Attach P to R as shown, with a further thumbtack positioned below. Stretch a rubber band between each thumbtack.

5 TIE a thread to U, and then around the winder spool. Tie a thread to M, run it over spool X in the jib, under and back over spool 3, and then through the screw eye. Glue the end to spool 3.

Winder and ratchet
To stop the jib from slipping, the jib winder is placed on a ratchet. A pivoting arm, P, is pulled against the slanting teeth on the winder wheel by a rubber band so that the winder can turn one way but not the other — unless you lift the ratchet arm.

Finishing Touches
Position the completed cradle on the base. The teeth from E should interlock with those of 5. Secure with G through F, E and A, placing thumbtacks at each end to fasten.

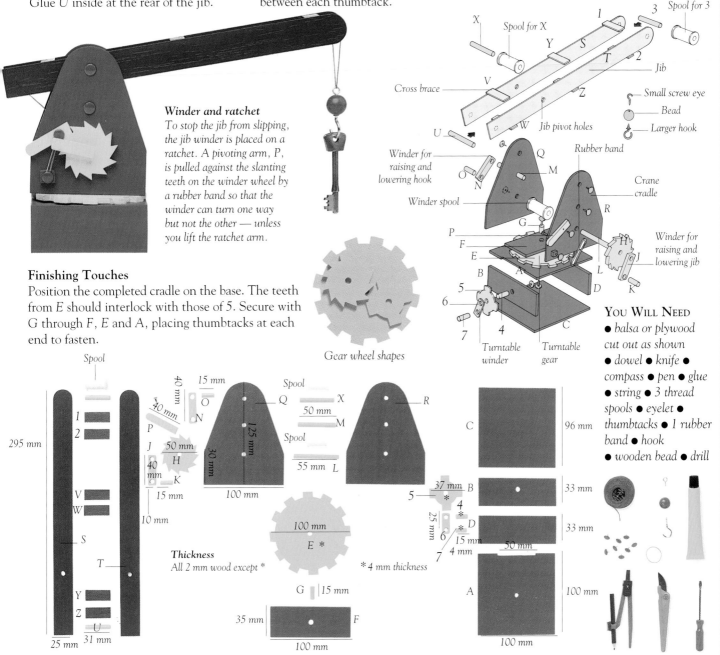

Gear wheel shapes

Cross brace

Spool for X

Jib

Small screw eye

Bead

Larger hook

Jib pivot holes

Winder for raising and lowering hook

Winder spool

Rubber band

Crane cradle

Winder for raising and lowering jib

Turntable winder

Turntable gear

You Will Need
- balsa or plywood cut out as shown
- dowel ● knife ●
- compass ● pen ● glue
- string ● 3 thread spools ● eyelet ●
- thumbtacks ● 1 rubber band ● hook
- wooden bead ● drill

Thickness
All 2 mm wood except *

* 4 mm thickness

Spool

295 mm

25 mm

31 mm

40 mm

40 mm

40 mm

15 mm

50 mm

40 mm

15 mm

10 mm

125 mm

30 mm

100 mm

50 mm

55 mm

100 mm

35 mm

100 mm

37 mm

25 mm

15 mm

4 mm

50 mm

96 mm

33 mm

33 mm

100 mm

100 mm

Stretching and squashing

WHEN SOMETHING IS PULLED or pushed, it does not always move. Sometimes it changes shape instead. You can mold modeling clay into any shape you want, and it will keep the new shape. Squeeze a balloon or soccer ball, however, and it jumps back to its original shape as soon as you stop squeezing. Similarly, if you stretch a rubber band, it snaps back to its original length as soon as you let go. Substances that bounce back like this are said to be "elastic." Rubber is elastic. So too are certain metals – – especially when drawn out into a wire then coiled into a spring. But elastic substances can only take so much. Once stretched or squashed beyond their limit, called the "elastic limit," they either break or stay out of shape for good.

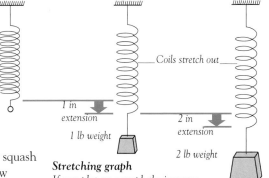

Squash ball?
When a tennis ball crashes into a racket, one side is squashed almost flat by the impact. Its elasticity makes it spring back into shape — but in squashing, the ball absorbs much of the energy of its momentum, and this energy helps speed it back again. The racket strings give a little to absorb some energy for a more controlled shot.

Spring extension
When a weight is hung on a spring, the spring's "extension" (increase in length) is proportional to the weight. Twice the weight stretches the spring twice as far. If you have a suitable spring, you can prove this for yourself. Spring balances work this way.

Coils stretch out

1 in extension

1 lb weight

2 in extension

2 lb weight

■ Hooke's law

Many elastic substances stretch or squash a regular amount, according to how much force is used. This is known as Hooke's law, after the English scientist Robert Hooke (1635-1703). But it only applies up to the elastic limit. Scientists talk of "stress" and "strain" in elasticity. By stress they mean the strength of the force in proportion to the area of the substance. Strain is how much the substance changes size. Hooke's law shows that the strain always varies directly with stress — that is, the amount a substance changes shape depends precisely on how strongly it is stretched or squashed.

Stretching graph
If you plot on a graph the increase in a spring's length for different weights, you get a straight line — up to the elastic limit. Rubber only obeys Hooke's law over a small range of weights, so its graph looks very different.

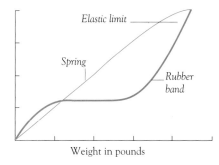

Elastic limit

Spring

Rubber band

Weight in pounds

Adult supervision is required. When something stretches, it absorbs energy from the stretching force. Exploit the energy in a stretched rubber band to turn a propeller and power a model plane. Parents of young children should do the cutting and drilling, and help with the assembly.

YOU WILL NEED
● *drill, pliers, knife ● ¹/₁₆ in thick balsa ● glue ● rubber band ● propeller from model shop ● wire bent to make propshaft and wheel struts ● beads ● button wheels*

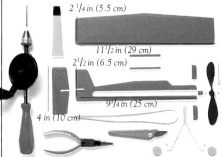

2 ¹/₄ in (5.5 cm)

11¹/₂ in (29 cm)

2¹/₂ in (6.5 cm)

9³/₄ in (25 cm)

4 in (10 cm)

1 CUT OUT the balsa shapes shown. Loop the band into the fuselage slot. Stick the tiny blocks into the nose. Drill a hole for the propshaft, and glue the nose strengtheners on each side.

2 GLUE SHORT STRIPS of balsa to the top of the fuselage to make a bigger wing joint. This joint will be under great strain if the plane crashes, so make sure it is strong.

3 GLUE THE WING to the top of the fuselage, with the straight edge facing the front. They must be exactly at right angles. Glue the tailplane in place.

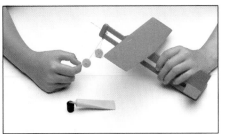

4 BEND THE WIRE for the wheel struts, and glue behind the nose. Strengthen with a loop of wire. Push on the wheels, and glue on beads to hold them in place.

5 SLIDE A BEAD and then the propeller onto the shaft. Bend the tip of the shaft over so that it sits in the groove on the propeller nose. Do not worry if it is loose — the rubber band will pull it tight.

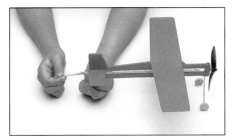

6 LOOP THE RUBBER BAND onto the hook on the propeller shaft. Stretch it, and hook it into the little notch on the back of the fuselage just below the tail. The plane is now ready for test gliding.

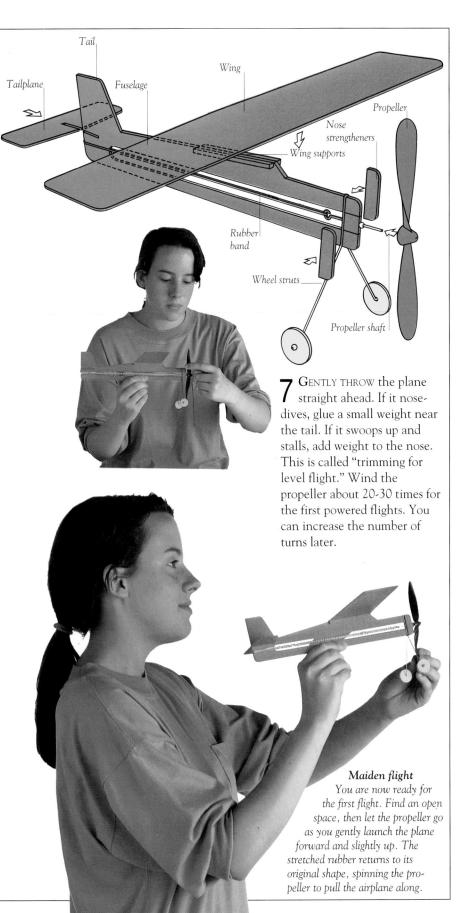

Tail

Tailplane

Fuselage

Wing

Propeller

Nose
strengtheners

Wing supports

Rubber
band

Wheel struts

Propeller shaft

7 GENTLY THROW the plane straight ahead. If it nose-dives, glue a small weight near the tail. If it swoops up and stalls, add weight to the nose. This is called "trimming for level flight." Wind the propeller about 20-30 times for the first powered flights. You can increase the number of turns later.

Maiden flight
You are now ready for the first flight. Find an open space, then let the propeller go as you gently launch the plane forward and slightly up. The stretched rubber returns to its original shape, spinning the propeller to pull the airplane along.

Light
■ AND ■
sound

MUCH OF OUR EXPERIENCE of the world comes through light and sound. Both are special forms of energy, each with its own particular qualities. The properties of light enable us to do everything from seeing ourselves in a mirror to creating three-dimensional holographic pictures. The properties of sound can create a crash of thunder or the charm of a flute.

When a distant firework explodes in a brilliant explosion of light and noise, you may see the cascading colors marginally before you hear the bang — because light travels through the air to you much faster than sound.

77

LIGHT

Without light, life would be impossible. We need light to see by. Plants grow by absorbing light and changing it to food. Indeed, light from the sun is the source of nearly all our energy and warmth. Yet light is just one kind of "electromagnetic radiation." There are other kinds, from radio waves to cosmic rays. Light is the only kind we can see.

In his book Opticks, *published in 1704, Isaac Newton suggested that a beam of light was a stream of tiny particles or corpuscles — starting a scientific controversy that was not resolved until early in the 20th century.*

If the wave theory of light is true, scientists in the 1800s asked, then how does light reach us from the stars through empty space?

The nature of light has intrigued people since the earliest times. The Ancient Greeks studied both reflection and refraction (the bending of light by glass or water, for example, pp. 84-85); Arab scholars in the Middle Ages, like Alhazen (c.965–1038), knew enough about refraction to make lenses for spectacles.

Yet it was not until the 17th century A.D. that there was any real progress in understanding what light really is. Then, at about the same time in the 1670s, the two greatest scientists of their age, Isaac Newton (1642–1727) and Christian Huygens (1629–95), came up with convincing, but completely contradictory, theories — opening a debate lasting over 200 years.

■ Corpuscles and waves

Newton suggested that a beam of light consisted of tiny particles, which he called "corpuscles," traveling at enormous speed. If light was corpuscles, he argued, this would explain why light travels in straight lines and casts sharp shadows. It also explained why mirrors reflected light — the corpuscles simply bounced off the glass like tennis balls off a wall. Refraction, Newton thought, might be caused by the corpuscles traveling faster in glass and water than in air. Most significantly of all, it explained why light could shine through a vacuum (p.102).

Newton showed white light is made from a spectrum of colors.

Huygens, by contrast, believed that light traveled as waves, like ripples on a pond. If this is so, Huygens showed, then it is easy to explain why white light is split up into a spectrum of colors as it is refracted through a prism, as Newton himself had discovered in 1666 (p. 94). Each color of light, Huygens suggested, has a different wavelength. If light travels *slower* through glass and water than through air, the amount of refraction — the amount light is bent — depends on the wavelength of the light. The shorter the wavelength, the more it is bent.

So scientists were presented with a dilemma. Huygens' wave theory explained the spectrum very well. But it did not

A pen seems to bend in a glass of water because light waves travel slower in water.

explain why light travels in straight lines and casts sharp shadows — sound waves and waves in the sea wash around objects in their path. And how could light travel through a vacuum in which there was nothing to wave? Newton's corpuscle theory, on the other hand, explained straight light rays and sharp shadows, but could not explain the spectrum.

For a century or more, the corpuscle theory seemed more popular, both because of Newton's name and because of the difficulty of explaining how waves could move through a vacuum. Then, in 1801, the English physicist Thomas Young (1773–1829) performed an experiment that, by demonstrating "interference," seemed to establish the wave theory beyond doubt (p. 100). Further discoveries during the 19th century reinforced the wave idea. But if light is waves,

Understanding how glass bent light helped Antony van Leeuwenhoek develop the microscope in the 17th century.

Electromagnetic spectrum
Visible light is only a small part of the vast range of electromagnetic radiation.

Radio waves

Television waves

Radar waves

Microwaves: used to heat food. In space they are echoes from the birth of the Universe

19th-century scientists wondered, how can light travel through a vacuum? Indeed, how can light reach us from the stars through billions of light-years of empty space? In answer, they came up with the idea of a special invisible material which they called luminiferous (light-bearing) ether.

■ Electromagnetism

Then, in the 1860s, the brilliant Scottish physicist James Clerk Maxwell (1831–79) made a crucial connection between light and electricity and magnetism. After ingeniously developing the idea of fields of electromagnetic force (p. 158), Maxwell showed that changes in the field happened at a speed identical to the speed of light. He then argued that light was just a small part of a wide range of electromagnetic waves — an idea confirmed when Hertz discovered radio waves in 1888. We now know that light is part of a huge spectrum of electromagnetic waves ranging from gamma rays to radio waves.

By the 1880s, it seemed almost certain that light was waves — and yet nagging doubts remained. These doubts were deepened in 1887 when American physicists Edward Morley (1838–1923) and Albert

Movie films record thousands of photographic images made by the effect of light on chemicals on the film.

Mixing every color makes white, as this spinning disc shows.

Colored bands on soap bubbles (p. 95) are caused by interference between light waves.

Michelson (1852–1931) proved there could be no such thing as luminiferous ether.

Then, in 1900, studies of the way heat was radiated from a hot object led the German physicist Max Planck (1858–1947) to a strange conclusion. Like light, heat radiation is a form of electromagnetic radiation, and Planck was trying to find a mathematical equation that would fit the pattern of radiation observed. But the only equation that worked was one that assumed radiation was emitted in tiny packets of energy or *quanta*, from the Latin for "how much?" Even Planck himself found this hard to accept, for it seemed to contradict 100 years of mounting evidence for waves. It was the genius of Albert Einstein (1879–1955) that brought out the real significance of the quantum theory.

■ Photons

In 1902, the German physicist Philipp Lenard discovered that when light struck certain metals, electrons were emitted — as if the impact of the light knocked electrons off atoms in the metal. This is how solar cells work.

The 19th-century physicist James Clerk Maxwell showed that light was just one form of a broad spectrum of electromagnetic radiation.

When physicists studied this "photo-electric effect," they were astonished to find that its intensity varied not with the intensity of the light, as expected, but only with the color of light.

Einstein showed that these mysterious results could easily be explained by thinking of light not as waves but as little packets of light energy, like Planck's quanta. These packets of energy were later called "photons." In a way, Einstein was reviving Newton's light corpuscles. But Einstein's photons are not particles like tiny balls at all; they can behave as waves also.

At first, many scientists were unhappy about the photon idea. It was hard to believe that light can be both a particle and a wave. It was harder still to believe that it is never both at the same time; sometimes it behaves as a particle and sometimes as a wave, depending on how you look at it. Yet the photon idea at last explained how light could travel through empty space and many other phenomena. Every experiment since has confirmed Einstein's theory, and the dual nature of light is now widely accepted.

Like ordinary lenses, Fresnel lenses focus light, but are robust enough for spotlights and car headlights.

Flesh is as transparent to X-rays as glass is to visible light. X-rays were discovered by the German physicist Wilhelm Roentgen (1845–1923) in 1895.

Infrared: heat rays

Visible light: red, orange, yellow, green, blue, indigo, violet

X-rays: flesh is transparent to X-rays

Gamma rays produced in nuclear reactions

Cosmic rays

Ultraviolet rays: tan the body

Light and shadow

LOOK AT THE BEAM of a flashlight piercing the darkness, and you can see that light rays travel in straight lines. If they hit an object in their path, they pass right through, bounce off, or are absorbed, just as a sponge absorbs water. Materials like glass let light through and are called transparent; substances that stop light passing are called opaque. When light hits an opaque object, it casts a shadow — an area where light does not reach. If you hold your hand under a bright desk lamp, you can see that there are two different kinds of shadow under your hand. In the middle is a very dark shadow. Around the edge is a narrow band of much lighter shadow. The dark shadow is called the umbra and can be seen wherever light rays are blocked off altogether. The lighter shadow, called the penumbra, forms where some of the light creeps around the edge of your hand.

Day and night are actually just light and shadow. The Earth spins on its axis as it travels around the Sun. So, at any time, half of the Earth faces the Sun, and there it is day. The other half of the Earth is in the Earth's shadow; here it is night.

The Diamond Ring effect
Seen just before and after a total solar eclipse, this is a sunbeam shining through a lunar valley.

A solar eclipse (right)
When the Moon blocks the Sun's rays, it casts a complete shadow (umbra) and a semi-shadow (penumbra) on the Earth. If you are in a part of the world where the umbra falls, you will experience a total solar eclipse. If you are in the penumbra, you will experience a partial eclipse.

■ An eclipse: the ultimate shadow

When the Earth lies directly between the Sun and the Moon, the Moon passes into the Earth's shadow and becomes dark. This is called a lunar eclipse. A solar eclipse takes place when the Sun, Moon, and Earth are in a direct line — a rare occurrence. The Moon, passing between the Sun and the Earth, casts a shadow over part of the Earth. The Moon's shadow hides the Sun's disc.

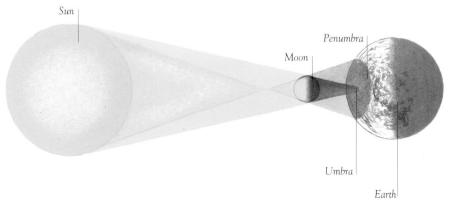

Sun

Penumbra

Moon

Umbra

Earth

Shadows short and long
In bright sunshine, stand in the same place and mark the position and length of your shadow every hour. What happens to the shadow as the hours go by? Try the experiment in summer and winter.

■ DISCOVERY ■
The sundial

AS THE SUN moves from east to west during the day, so the shadows it casts move too. The Babylonians learned to tell the time by this over 4,000 years ago, making the first shadow clock, or sundial. Every sundial has an upright called a gnomon, and the gnomon's shadow shows the time on a flat dial marked in hours.

EXPERIMENT
Making a sundial

YOU WILL NEED
● *flowerpot* ● *stick (twice the height of the flowerpot)*
● *black marker pen or crayon*

On sunny days you can tell the time with a sundial made from a flowerpot and a stick. The stick forms the gnomon; the pot is the dial.

1 Turn the flowerpot upside down, and push the stick through the base hole into the ground.

2 Place the pot in the sun. Mark the position of the shadow cast by the stick every hour. Each time it is sunny, you will be able to tell the time from the marks on your sundial.

1 A SIMULATION OF A SOLAR ECLIPSE using a globe (Earth), lamp (Sun), and a small ball (Moon). As the Moon rotates around the earth it passes in front of the Sun and partly obscures it.

2 WHEN THE EARTH, MOON, and Sun are in a direct line, a total solar eclipse occurs. For a short time, day becomes night in the parts of the world where the shadow is cast.

Reflection

WITHOUT LIGHT we could not see the world around us. Yet only a few objects, such as the Sun, are naturally luminous — that is, give out light. We see most things by the light bouncing off them. When you see a chair in daylight, what you really see is light from the sun reflected from the chair. Just how much light is reflected depends on the surface. A smooth white surface, for instance, reflects more light than a dark rough one. Mirrors have such smooth, shiny surfaces that light bounces off in exactly the same pattern as it arrives, reflecting a complete picture or image of any object. In a flat mirror, you see an image the same size as the object. In a curved mirror, the image may be bigger or smaller — as you can see by looking at a spoon. A convex (bowed out) mirror gives a smaller image. A concave (dished) mirror gives a larger image of a nearby object, and a smaller, upside-down image of a distant object.

Infinity mirrors
Set up two facing mirrors so that they are almost parallel. Position yourself between them. You will see a series of images of yourself as light bounces to and fro between the mirrors. Because each mirror absorbs some of the light that hits it, the last images in the series disappear.

Reflecting power

Different surfaces reflect light in varying degrees. White surfaces reflect light well. Dark and matte black surfaces reflect very little light, absorbing most of it. In this simple experiment you can test how dark and light surfaces reflect light. Use a flashlight as the light source and a piece of white cardboard as a screen.

YOU WILL NEED
● *flashlight* ● *small flat mirror* ● *white cardboard* ● *matte black cardboard*

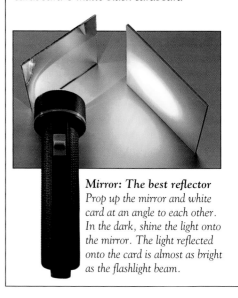

Mirror: The best reflector
Prop up the mirror and white card at an angle to each other. In the dark, shine the light onto the mirror. The light reflected onto the card is almost as bright as the flashlight beam.

Making a periscope

Adult supervision is suggested.
A submarine captain uses a periscope to observe ships on the surface. This is a long tube with a prism (a triangular piece of glass) at either end that reflects light down the tube. A simple periscope works with parallel sloping mirrors instead of prisms. It enables you to see things that are not in your direct line of vision.

YOU WILL NEED
● *stiff black card, 2 ft x 12 1/2 in (60 x 32 cm)*
● *2 small flat mirrors, 4 x 3 in (10 x 7.5 cm)*
● *protractor* ● *colored gummed paper*

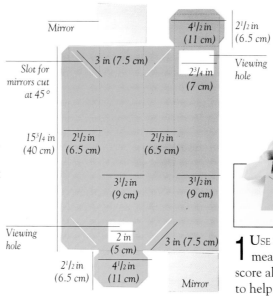

Mirror
4 1/2 in (11 cm)
2 1/2 in (6.5 cm)
Viewing hole
Slot for mirrors cut at 45°
3 in (7.5 cm)
2 3/4 in (7 cm)
15 3/4 in (40 cm)
2 1/2 in (6.5 cm)
2 1/2 in (6.5 cm)
3 1/2 in (9 cm)
3 1/2 in (9 cm)
Viewing hole
2 in (5 cm)
3 in (7.5 cm)
2 1/2 in (6.5 cm)
4 1/2 in (11 cm)
Mirror

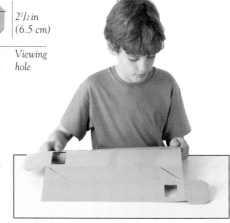

1 USE THE TEMPLATE (left) as a guide to measure the shape you need. Lightly score along the lines. Use the protractor to help you cut the angled slits and holes.

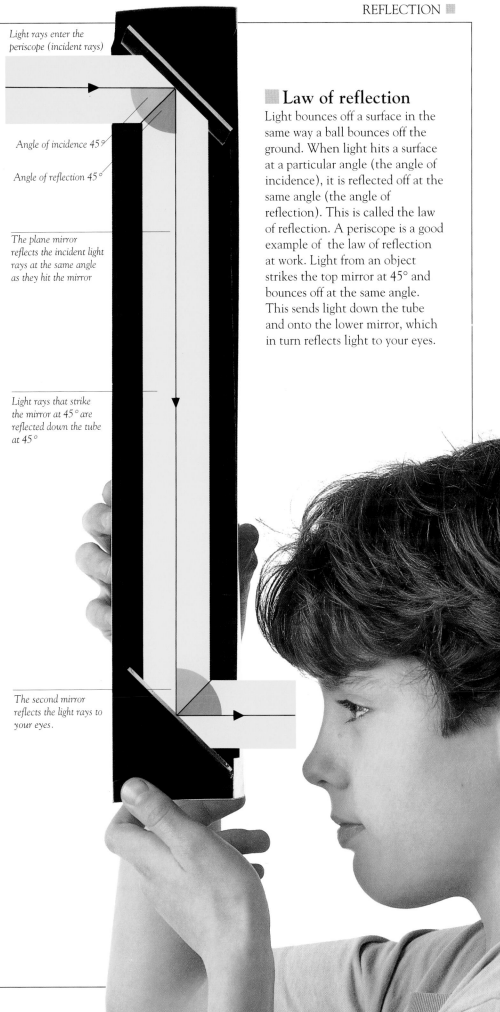

Light rays enter the periscope (incident rays)

Angle of incidence 45°

Angle of reflection 45°

The plane mirror reflects the incident light rays at the same angle as they hit the mirror

Light rays that strike the mirror at 45° are reflected down the tube at 45°

The second mirror reflects the light rays to your eyes.

Law of reflection

Light bounces off a surface in the same way a ball bounces off the ground. When light hits a surface at a particular angle (the angle of incidence), it is reflected off at the same angle (the angle of reflection). This is called the law of reflection. A periscope is a good example of the law of reflection at work. Light from an object strikes the top mirror at 45° and bounces off at the same angle. This sends light down the tube and onto the lower mirror, which in turn reflects light to your eyes.

Testing white
Replace the mirror with a white card. The flashlight beam is reflected well, but not so brightly as with the mirror.

Testing black
Now continue the experiment with a piece of black cardboard. You will see that it reflects almost no light. The color black absorbs almost all the light that hits it.

2 FOLD THE PANELS to make the tube. Fold and tuck in the flaps, and join the edges with adhesive tape.

3 SLIDE THE MIRRORS through the angled slits so that they are parallel and facing each other. Fix the mirrors in position with adhesive tape.

Refraction

HAVE YOU EVER NOTICED that a swimming pool always looks shallower than it is, or that your legs appear to bend in the middle when you stand in the bathtub? This happens because light rays bend when they pass from one transparent substance to another — in this case, from water to air. This effect is called refraction. Refraction occurs because light travels at different speeds through transparent materials. Light moves at about 186,000 miles per second (300,000 kilometers per second) through air. When it enters glass or water, it slows down, which makes it change direction. Refraction can be very useful. Lenses are specially shaped pieces of glass that refract light in a precise way. For instance, a spectacle lens corrects the way light enters a defective eye to give a clear image.

The "bent" pen
Rest a pen at an angle in a glass of water, and then stand back from the glass. The pen appears to bend or split into two sections. This is because light refracts as it passes from water into air.

■ Types of lens

Lenses are used in almost every optical device, from cameras to telescopes. There are two basic types of lens: convex and concave. A convex lens is thicker in the middle than at the edges. It converges light rays passing through it toward a point called the focus, and can make objects look larger. A concave lens is thinner in the middle than at the edges. It spreads the light rays passing through it and makes objects look smaller.

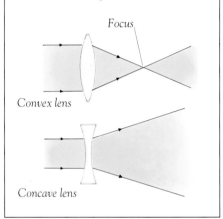

Focus

Convex lens

Concave lens

EXPERIMENT
Seeing light bend

In this experiment, you can see how water refracts light rays. The angle of refraction (the angle at which light bends) depends on the angle at which the rays hit the bottle. Moving the bottle around will give different angles of refraction.

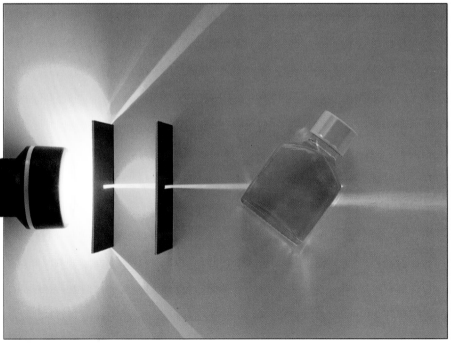

YOU WILL NEED
● *square-sided bottle with cap* ● *2 pieces of white card (4 x 3 in/10 x 8 cm)* ● *sheet of white paper* ● *flashlight* ● *a few drops of milk*

1 FILL THE BOTTLE with water, add enough milk to cloud the water, and replace the cap. Cut narrow slits (0.06 in wide) in the cards; put them in front of the flashlight.

2 LAY THE BOTTLE on the paper at an angle. Darken the room, and shine the flashlight through the slits so that a thin beam of light hits the bottle.

3 THE MILK in the water helps you see light bend inside the bottle. Notice how the beam bends back when it comes out of the other side of the bottle.

Making a microscope

A magnifying glass (a single convex lens) is a simple microscope. However, most microscopes have several lenses for greater magnification. In the simplest microscopes, there are two lenses. Closest to the object being studied is the "object" lens. The other lens, called the "eyepiece" lens, is the lens you look through. The eyepiece lens is usually much bigger. You can make a very basic microscope with a magnifying glass as the eyepiece lens and a powerful convex lens (such as an old camera lens or a photographer's loupe).

YOU WILL NEED
- *a magnifying glass* ● *a strong convex lens*
- *object to be studied (such as a flower)*

1 Hold the magnifying glass close to your eye, and place the convex lens just above the object, which should be about 2 ft (60 cm) from your eye.

2 Adjust the position of the lenses until an enlarged image of the object appears in focus.

■ DISCOVERY ■
Hooke's Microscope

ROBERT HOOKE, the 17th-century English physicist, wrote that with a microscope we can "peep in at the windows" of nature. A pioneer of lens development, Hooke constructed a compound (multiple-lens) microscope to study living things in minute detail. The specimens were illuminated by an oil lamp that had a circular reflector and a lens (called a condenser) to concentrate the light. With the aid of his microscope, Hooke discovered that plants and animals are made up of tiny cells. His book *Micrographia*, published in 1665, includes his detailed drawings of insects under the microscope.

Hooke's drawing of a compound microscope from his book Micrographia.

Experimenting with light

SCIENTISTS have always been intrigued by the behavior of light. As a result of the discoveries of people such as the great Isaac Newton, we know that certain laws govern how light rays are reflected, refracted, and focused. Simple experiments with light can show you how these laws work.

EXPERIMENT
The light-ray house

Parental supervision required.
See how light is refracted when it passes through transparent objects such as bottles and lenses. Experiment with mirrors and other shiny objects to learn about reflection.

YOU WILL NEED
● *black cardboard, 12 x 12 in (30 x 30 cm)* ● *white cardboard, 15 ¹/₂ x 15 ¹/₂ in (40 x 40 cm)* ● *craft knife* ● *adhesive tape or glue* ● *battery* ● *2 lengths of electrical wire with alligator clips* ● *bulb and holder* ● *glass bottle* ● *mirror* ● *glass tumbler* ● *convex lens* ● *small comb*

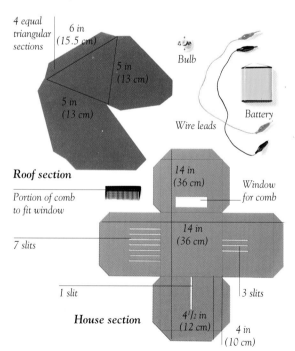

4 equal triangular sections

6 in (15.5 cm)

5 in (13 cm)

5 in (13 cm)

Bulb

Wire leads

Battery

Roof section

Portion of comb to fit window

7 slits

14 in (36 cm)

Window for comb

14 in (36 cm)

1 slit

3 slits

House section

4¹/₂ in (12 cm)

4 in (10 cm)

1 CUT OUT THE HOUSE SECTION using the template above as a guide. Cut narrow slits in three sides as shown. Cut a window in the fourth side. Score and fold along the solid lines to make the box.

Bottle lens
Fill the small bottle with water, and stand it in the path of two light rays. The bottle acts as a convex lens, refracting and converging the rays that strike it. The rays of light meet at a point called the focus or focal point. Try bottles of different diameters to see how this affects where they focus the light.

Convex lens
Position the lens in line with the light rays from the comb window. Notice that the lens focuses light farther away than the bottle (above). This is because the lens is less curved than the bottle. The greater a lens's curvature, the stronger it is. Replace the convex lens with a concave lens. See how it makes light rays spread out.

Mirror
Place the small mirror at an angle to the single ray of light. Repeat this with different angles, and use a protractor to measure the incident and reflected angles each time. What is the relationship between the two angles? Now experiment with curved reflectors such as a bottle wrapped in aluminum foil.

Tumbler lens
Fill the wide-bottomed glass with water, and position it in front of the three slits. Note the position of the focal point. Now empty the glass and refill it with cooking oil. The focal point is much closer to the glass. This is because oil refracts light more than water.

■ Setting up the ray house

Place the house in the middle of the sheet of white cardboard, and position your reflectors and refractors around the four walls. In a dark room you will be able to see how the light rays change direction when they strike each of the objects.

2 TAPE THE EDGES of the box in place. Now fold the roof section along the lines as shown above.

3 TAPE THE SMALL COMB behind the window. Place the battery, the bulb, and the bulb holder inside the box.

4 CONNECT the battery terminals to the screws on either side of the bulb holder. Screw in the bulb so that it lights up. Place the roof on the house.

Images and illusions

THE EYE IS a fascinating optical instrument. It detects colors, focuses on its subject automatically, and rapidly adapts to the brightness of light. About 150 years ago, the pioneers of photography succeeded in capturing the images that they saw. They made cameras that worked like primitive eyes and invented the photographic process to record the image. Today, cameras are very sophisticated. However, only the most modern automatic cameras have all the automatic features of the human eye.

Camera obscura
The first camera, the camera obscura, used a lens to project an image of the subject onto a screen.

■ DISCOVERY ■
Early photography

DURING THE 16TH CENTURY, people learned how to project an image of an object using a camera obscura, but until the early 19th century, there was no way of recording the image. In 1826, a French inventor, Joseph Niepce (1765–1833), took the first photograph. He coated a metal plate with light-sensitive chemicals and projected an image onto it using a camera obscura. It took about eight hours to record the picture, which was fuzzy and indistinct. In 1837, Frenchman Louis Daguerre (1787–1851) improved Niepce's system, which became know as the "Daguerreotype" process. In 1839, Englishman William Fox Talbot (1800–77) invented a method of taking pictures onto a negative (in which light and dark are reversed) from which countless positive prints could be made. Prints are made the same way today.

The eye and the camera

The camera is similar to the human eye in many ways. Both have essentially three components: a hole that opens and closes to let in the right amount of light, a lens for focusing the light, and a device that records the image. In the case of the eye, a band of muscle called the iris opens and closes the pupil, which is a hole at the front of the eye. Behind the pupil is the lens. The lens is surrounded by muscles that change its shape so that it can focus on near and distant objects. The lens focuses light on cells in the retina, which lines the back of the eye. When light hits the cells, they send messages to the brain, which decodes them into an image. A camera has a variable-size hole called the aperture and a shutter that opens briefly to allow light into the camera. Both the speed of the shutter and the size of the aperture control the amount of light that hits the film. A series of lenses move in and out to focus the light on to the film, which lies at the back of the camera. Once the film is developed and fixed, a negative image appears on the film. This image is printed onto photographic paper which is also developed and fixed to produce the final picture.

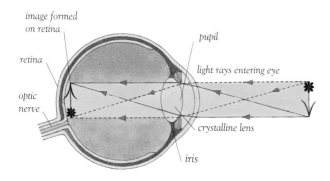

The eye
The eye focuses an upside-down image of an object onto the retina. At a very early stage in a baby's development, the brain learns to interpret the image so that it appears the right way up.

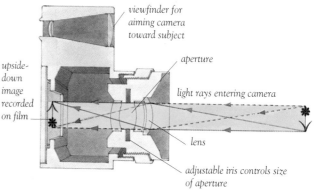

The lens camera
The image formed on the film is a negative. This means that once it is developed, dark parts of the subject appear light. When the negative is printed on to photographic paper, a normal (positive) image is formed.

Optical illusions

Our brains can be tricked by the things we see. This produces optical illusions. For instance, if you walk along a street on a very hot day, you sometimes see a mirage. An image of a pool of water appears on the road ahead and recedes as you approach it. A mirage is caused by light from the sky reflecting off a layer of hot air near the ground. The brain wrongly interprets the blue color of the sky as water.

The impossible tribar

This is an example of another kind of optical illusion. The structure above looks normal and does not confuse the eye. But when it is redrawn so that all the pieces join up, our eyes become confused.

The tribar looks as though it is made of three right-angled sections that are joined to form a triangle. Your brain knows that this could not happen in a three-dimensional object, so the drawing seems to show an impossible shape.

Moving pictures

Moving pictures such as television and movie films rely on an optical illusion. When you see two pictures in quick succession, the image appears to move because your eyes have not been able to tell that the picture has changed. When you look at a picture, the brain retains the image for about one-tenth of a second after it has gone. This is called persistence of vision. So, if you are shown more than 10 still pictures a second, your brain will join the separate images into a series of moving pictures. You can see how this works by drawing a series of pictures of a person, one on each page of a notebook, with each picture slightly different from the last. When you flick through the pictures quickly enough, the person seems to move. Motion pictures show 24 photographic frames each second. A projector shines a strong light through the film, and a lens focuses the image onto a screen. The film runs through the projector in rapid jumps, each bringing a new frame into view. A television picture consists of a series of horizontal lines, drawn one after another down the screen. The lines are redrawn many times each second to make a new picture.

Figures in motion

Throughout the 19th century, many inventors worked on ways of making motion pictures. In 1877, Eadweard Muybridge (1830–1904) became the first person to photograph movement. He captured pictures of people and animals in motion in still pictures taken by a line of cameras. He then mounted the images onto a rotating glass disc, which he viewed through a slotted disc that rotated in the opposite direction. He even shot galloping horses with cameras fired by the horses themselves as they broke through a series of strings.

Projector

Some Victorian households had "moving theaters" for entertainment. Frenchman Émile Reynaud (1844–1918) developed the first of these "wonder cylinders," which projected sequences of photographs or drawings as moving pictures.

Recording images

PHOTOGRAPHY IS SIMPLE in principle. All you need are a camera and light-sensitive material. A camera can be just a lightproof box with a tiny hole instead of a lens. The light-sensitive material may be film or paper coated with an "emulsion" dotted with silver salt crystals. When the hole is opened to take a picture, crystals hit by light are altered so that when the film is developed in chemicals, they change into silver. After "fixing" in more chemicals to remove unchanged silver salts, the image is preserved in the pattern of silver grains.

Outer box

Inner box

Pinhole

Sliding shutter

3³/₄ in (9.5 cm)

3¹/₄ in (8 cm)
3¹/₄ in (8 cm)

4¹/₂ in (11 cm)
2¹/₂ in (6.5 cm)

3¹/₂ in (9 cm)

3¹/₄ in (8 cm)

4¹/₂ in (11 cm)
3¹/₄ in (8 cm)

3¹/₂ in (9 cm)

¹/₂ in (1 cm)
1¹/₄ in (3 cm)
¹/₄ in (0.5 cm)
³/₄ in (2 cm)
4 in (10 cm)

EXPERIMENT
Pinhole camera

YOU WILL NEED

For the camera: existing box or can, or stiff black card to make box as shown here ● black carpet tape ● aluminum foil ● small pin
For processing: ● water ● 3 plastic trays ● measuring cup ● developer, fixer, photographic paper, and red "safelight" (from photographic supplier) ● clear Lucite ● thick black (lightproof) plastic ● flashlight ● plastic tweezers

1 CUT OUT THE BOX shapes in blue and yellow above from card, and also the red shutter. Score the lines, fold back sharply, and glue tabs to make two boxes. Stick carpet tape down the inside of all joints to make the boxes lightproof.

2 MAKE SURE THE YELLOW BOX fits snugly into the blue outer box, open end first. Paint matte black inside. Cut a small hole in the center of the base of the yellow box, and stick aluminum foil firmly over it on the inside.

3 PRICK A HOLE through the foil with a pin, then use emery paper to sand off the jagged edges. Make a small hole in the shutter in line with the pinhole.

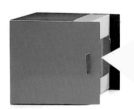

4 IN YOUR DARKROOM (see right), unwrap the photographic paper, take out one sheet, and close the wrapper again. Place the sheet shiny side up in the base of the blue box. Slot the yellow box in on top, make sure the shutter is "closed," and seal all joints with carpet tape. Put the camera in a thick bag, and turn on the lights.

5 CHOOSE YOUR SUBJECT — it must be in bright daylight — and put your camera on a flat surface pointing at it. Then slide the shutter to open the pinhole and expose the photo. Exposure times can vary widely, but for an average subject on a bright, cloudy day try about four minutes. Close the shutter, put the camera in the bag again, and return to your darkroom.

Darkroom and processing

Adult help is needed for darkroom work.

A kitchen or bathroom makes a good darkroom, because you need running water. Make the room completely lightproof by taping thick black plastic over the windows and hanging a thick blanket over the door frame. Check that there are no light leaks, waiting for a few minutes while your eyes adjust to the dark. Connect a shop-bought red safelight, or fit a 15-watt bulb in a desklamp, and cover the mouth and all holes with red safelight filter. Mix the developing chemicals in ordinary light, following the maker's instructions. Set three trays in a row on a waterproof surface, in the order shown above. Put the tweezers, a sharp knife, and a flashlight at hand where you can find them easily. Switch on the safelight, and seal yourself in your darkroom.

3 plastic trays

Developer   *Water*   *Fixer*

Developer   *Fixer*

Clear Lucite   *Photographic paper*

Tweezers   *Measuring cup*   *Flashlight*   *Red safelight filter*

1 CAREFULLY SLICE through the tape to separate the camera boxes — for later shots, lay tape over the old tape, then peel off for processing. Take out the paper, and slide shiny side up into the developer. Make sure it is fully immersed. A negative image should appear. Let darken for 2-3 mins, then transfer it with tweezers to the water for 1 min. Next, transfer it to the fixer for 5 mins, then wash well in running water and hang to dry.

After the first processing stage you get a negative (with dark and light reversed). This must now be "printed."

2 TO PRINT, place the negative face down on photographic paper (shiny side up), and cover it with a square of clear Lucite to keep it in position.

3 SHINE A FLASHLIGHT onto the negative for about 10 seconds. Remove the photographic paper, and repeat the developing process: developer, water, fixer, wash, dry.

The final print — you may have to experiment a few times with exposure and processing time until you get a good picture.

Moving pictures

DURING THE 19TH CENTURY, the closest thing to watching a video would have been playing with a toy called a zoetrope. It consisted of a slotted drum lined with a strip of drawings. As the drum rotated, the pictures appeared to move. As with today's movies, zoetropes rely on persistence of vision. When a slit passes in front of the eye, one picture is seen for a fraction of a second. Before that image fades in the brain, the drum moves on and then another picture comes into view. If the drum spins fast enough, the pictures merge and seem to move.

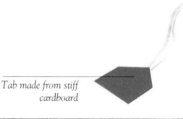

String to rotate drum

Viewing window into which base of rotating drum is fitted (see exploded diagram)

The zoetrope (left)
The zoetrope on these pages is based on the original model. The word zoetrope comes from the Greek and means "wheel of life."

Tab made from stiff cardboard

Juggling clown
The pictures for the zoetrope show different stages of movement, as with this juggling clown.

①

②

EXPERIMENT
Making a zoetrope

Parental supervision required.
Constructing your own zoetrope involves some careful and intricate assembling, giving you a fine working model to keep. The exploded diagram (below right) shows how the various pieces are fitted together. For the moving pictures, trace the clowns below or draw your own picture sequence.

YOU WILL NEED
- *stiff colored card 20 x 28 in (51 x 71 cm)* • *paper glue* • *small dowel or toothpick* • *string or thread* • *tape*
- *colored paper for decorations*

1 Curve piece B, and glue it to piece F. Mark the center points of pieces E, K, and J.

2 Glue together J, G, and K. Now insert the dowel with a blob of glue on the end and plenty of glue around the hole in piece K. Glue J to the underside of F, aligning the center markings.

3 Fold D and glue it to C. Curve H and join it to C. Glue E in position. Slot the drum into place by fitting the dowel into piece D.

4 Assemble the viewing window by gluing A and L together, then inserting I.

5 Tape a length of string to the drum spindle. Wind this around the spindle, but leave enough to thread through the viewing window. Pull the string and watch the pictures move!

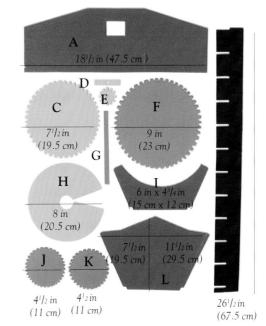

A — 18½ in (47.5 cm)
D
E
C — 7½ in (19.5 cm)
F — 9 in (23 cm)
G
H — 8 in (20.5 cm)
I — 6 in x 4¾ in (15 cm x 12 cm)
J — 4½ in (11 cm)
K — 4½ in (11 cm)
L — 7½ in (19.5 cm) 11½ in (29.5 cm)
26½ in (67.5 cm)

Circular strip depicting images of juggler.

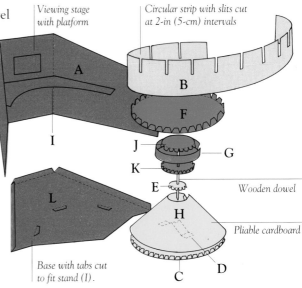

Viewing stage with platform
Circular strip with slits cut at 2-in (5-cm) intervals
A
B
F
I
J
G
K
E — Wooden dowel
H — Pliable cardboard
L
C — D
Base with tabs cut to fit stand (I).

Attach ② here

The mechanism (below)
When pulled, the string wrapped around the spindle spins the drum.

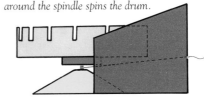

Attach ① here

The color of light

LIGHT COMES IN TINY WAVES. But not all waves are the same size. Some waves are short; some are long. When we see colors, we are seeing light of different wavelengths. Sunlight appears to be colorless — it is called white light — but it is really a mixture of many colors, or wavelengths, of light. You can see some of these colors when a rainbow appears in the sky after rain. As sunlight is reflected through raindrops it is bent, or refracted. Light with long wavelengths, like red, is bent more than light with short wavelengths, like violet; so the colors fan out as they re-emerge from the raindrop.

Sun

Sun's rays

Refraction

Reflection

Refraction

Raindrops

Red ray

Green ray

Observer's eye

Violet ray

Rays from the rain
Refraction and reflection of the Sun's white light rays in raindrops cause colored rays to spread out from the raindrops at different angles. Rays coming from raindrops in different parts of the sky enter the eye, so we see a curved band of colors. These colors — red, orange, yellow, green, blue, indigo, and violet — are called the spectrum.

■ DISCOVERY ■
Newton's color spectrum

SOME 300 YEARS AGO, the English physicist Isaac Newton directed a beam of sunlight through a slit into a darkened room and onto a triangular glass prism. The prism refracted the white light so that it fanned out into a spectrum of colors. When Newton placed a second prism in the spectrum, the light rays combined again to form white. Newton had proved that white light is made up of different colors. He published his results in his book *Opticks* in 1704.

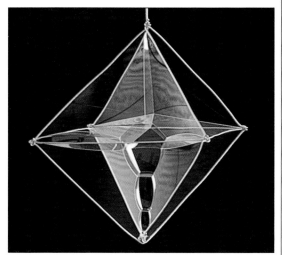

Bend the wire into simple shapes to make two-dimensional bubbles (soapy films) of figures or animals.

Producing the spectrum on paper

On a sunny day, you can see a spectrum by conducting an experiment that is similar to Newton's.

YOU WILL NEED
● *straight-sided tumbler* ● *piece of card with* $^1/_2$*-in (1-cm) slit* ● *sheet of white paper*
● *adhesive tape*

1 Fill a glass with water, and tape the card on at the slit (as shown).

2 Place the white paper close to a window, and stand the glass on it. Sunlight passing through the slit is refracted by the water in the glass. This produces a spectrum on the paper.

A geometric wire shape such as this octahedron produces a three-dimensional bubble.

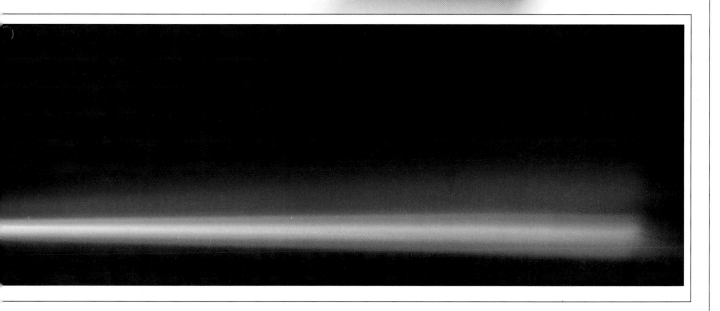

Mixing colors

HAVE YOU EVER WONDERED how thousands of colors appear in a television picture? In fact, all of them are made by mixing just three colors: red, green, and blue. These are called the primary colors. When red, green, and blue lights are mixed together, they produce white light; varying the proportions of the three colors produces all other colors. Mixing lights in this way is called color addition. However, we see most things only by reflected light, and they get their color by a process called color subtraction. An object absorbs, or subtracts, certain colors of the light falling on it and reflects the rest. It is these reflected colors that give the object its color.

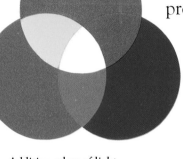

Additive colors of light
The primary colors of light are red, green, and blue. When mixed together, they make white light; when mixed in pairs, they produce the secondary colors yellow, magenta, and cyan.

■ Stage lighting

Colored lights are used to produce stunning effects at the theater. For example, changing the lighting from bright yellow to cold blue changes the mood on stage from a sunny day into an eerie night. Lighting technicians use powerful white spotlights and a range of colored filters to achieve these effects. A filter is a colored film that allows light of only one color to pass through it. Covering the spotlights with different filters produces colored light, and these lights can be mixed to produce new colors. Make your own colored filters from pieces of colored cellophane, and experiment with the principles of theater lighting.

Moody blues (right)
Good stage lighting enhances a production by illuminating the scene in warm or cold, hard or soft colors. A nighttime city effect is created here with mainly blue light and some magenta.

EXPERIMENT
Making a color viewing box

The color of an object depends on the color of light falling on it. A tomato looks red because it reflects light from the red part of the spectrum and absorbs the other colors. However, in a beam of light that does not contain red, the tomato looks black because there is no red light to reflect. A white object reflects all colors of light equally, so it appears to be the same color as the light shining on it. Test this color theory for yourself.

YOU WILL NEED
● *shoe box with a large rectangular hole cut in the lid and small hole in one end* ● *red and green cellophane* ● *scissors* ● *flashlight* ● *green apple* ● *banana* ● *playing card (heart or diamond)*

Seeing green
Tape the green cellophane under the lid of the box. Place the fruit and card inside, and position the flashlight.

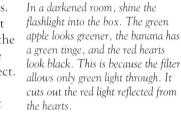

In a darkened room, shine the flashlight into the box. The green apple looks greener, the banana has a green tinge, and the red hearts look black. This is because the filter allows only green light through. It cuts out the red light reflected from the hearts.

Seeing red
Replace green cellophane with red, and shine the flashlight into the box.

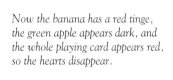

Now the banana has a red tinge, the green apple appears dark, and the whole playing card appears red, so the hearts disappear.

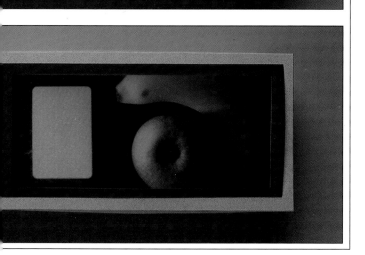

EXPERIMENT
Turning colors white

This experiment shows that white light is a mixture of colors. Each segment of the disc reflects light of a different color. When the disc spins, your eyes mix the colors to form white.

YOU WILL NEED
● *white cardboard* ● *scissors* ● *water-based paints (red, orange, yellow, violet, green, indigo, cobalt blue)* ● *paint-brushes* ● *sharpened pencil* ● *protractor*

1 Cut out the disc, and divide it into seven sections with a protractor. Paint the seven colors of the spectrum onto the disc.

2 Make a hole in the center of the disc to fit the pencil. Spin the wheel on the pointed end of the pencil, and watch the colors disappear!

Spinning wheel
When the disc spins rapidly, your eyes cannot distinguish the separate segments, so the colors merge. If perfectly pure colors were used in the correct balance, the disc would appear white. In fact, this is almost impossible. When you spin the disc, it will probably look pale gray.

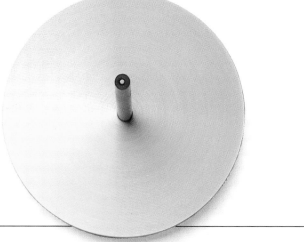

Pigments and paints

YOU MIGHT THINK that colors in photographs and movies are made in much the same way as colors on a television screen — by *adding* red, green, and blue light in the right proportions (p. 96). In fact, they are made in exactly the opposite way. The film is coated in dyes that, like filters, *subtract* or absorb red, green, or blue light from white light in the right proportions. Where you see red on the film, for example, there are dyes that absorb green and blue light from the light from the film projector, leaving just red. Not only films, but books, paintings, and any kind of printing use dyes and pigments to create colors in the same way.

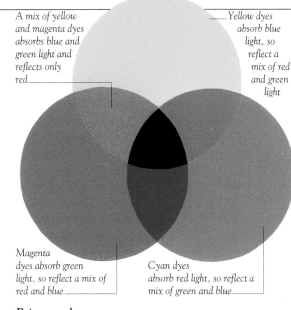

A mix of yellow and magenta dyes absorbs blue and green light and reflects only red

Yellow dyes absorb blue light, so reflect a mix of red and green light

Magenta dyes absorb green light, so reflect a mix of red and blue

Cyan dyes absorb red light, so reflect a mix of green and blue

Primary colors
Just as you can add the three primary colors of light — red, green, and blue — together to make any other color, so you can mix three basic colors of paints, dyes, or pigments together to make any other color. But primary pigments work by absorbing colors. So they are not red, green, and blue but yellow, magenta, and cyan pigments, each of which absorbs one of the primary colors of light and reflects the other two.

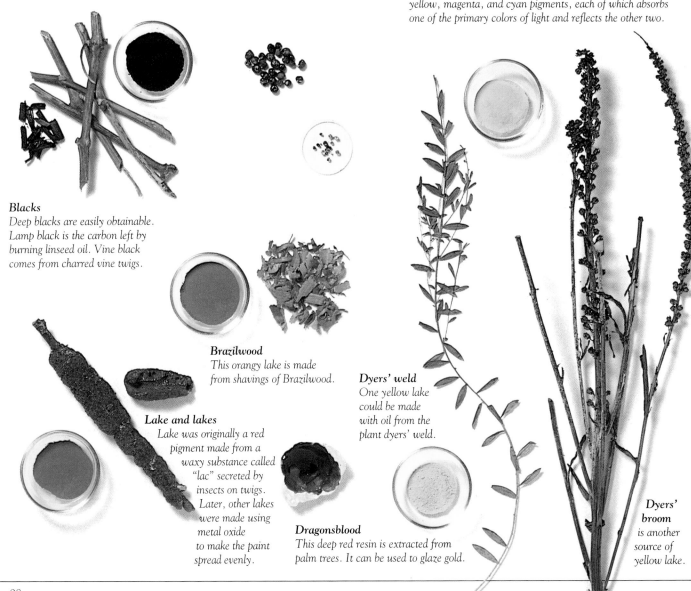

Blacks
Deep blacks are easily obtainable. Lamp black is the carbon left by burning linseed oil. Vine black comes from charred vine twigs.

Brazilwood
This orangy lake is made from shavings of Brazilwood.

Dyers' weld
One yellow lake could be made with oil from the plant dyers' weld.

Lake and lakes
Lake was originally a red pigment made from a waxy substance called "lac" secreted by insects on twigs. Later, other lakes were made using metal oxide to make the paint spread evenly.

Dragonsblood
This deep red resin is extracted from palm trees. It can be used to glaze gold.

Dyers' broom
is another source of yellow lake.

Mortar and pestle
In the Middle Ages, every artist had a mortar and pestle for grinding pigments from minerals.

The artist's palette

Pigments were used as long ago as the Stone Age, 12,000–30,000 years ago, when cave paintings were made with paints ground from colored earths and clays, such as iron oxide and yellow ocher — often mixed in animal fat to make them waterproof. The range of pigments the neolithic artists used was fairly limited, consisting mainly of browns and reds. But through the centuries, more and more natural pigments were found, and by the Middle Ages, artists had a wide range of natural pigments to paint with, and clothmakers had dyes of many hues — from rich blues made from lapis lazuli to vivid greens from malachite. Some pigments were ground from minerals and earths; others came from plants and animals. Growing trade links between Europe and the Orient enlarged the range still further. Only in 1856 was the first synthetic dye, the purplish "mauvein," made. Since then, hundreds of artificial dyes and pigments have been created.

Renaissance paints

Paintings of the early Italian Renaissance (14th and 15th century) are usually in emulsions of egg tempera — first used by the ancient Egyptians. Later, the Italian masters often painted in oils, using ground pigments suspended in linseed oil.

Red earth
One of the oldest natural pigments, this reddish brown ocher is ground from iron oxide found in the earth.

Yellow earth
Like red earth, this ocher is ground from iron oxide. The yellow comes from mixing with water in the ground.

Malachite
This greenish pigment comes from a copper ore called malachite (copper carbonate-hydroxide)

Red lead
This pigment, manufactured by heating lead oxide to 400°C, is now rarely used because it is poisonous.

Binder
To coat evenly, a pigment is mixed with a binder. Egg tempera, from egg yolks and water, gives a clear look.

Green earth
"Terre verte" comes from a greenish clay containing the mineral viridian.

Ultramarine
This highly prized intense blue pigment comes from the semiprecious stone lapis lazuli.

Azure
This turquoise pigment comes from a copper ore called azurite (copper carbonate).

Vermilion
This bright red — used in prehistoric China — is made from cinnabar (mercuric sulfide) from volcanic rocks.

White lead
Artists need white as much as any other color. This is the traditional artist's white.

Waves of light

TWO CENTURIES AGO, there were two conflicting theories of light. Some scientists, following Isaac Newton, believed light is made up of tiny particles, or "corpuscles." Others, such as Thomas Young, thought light consists of waves. Young provided powerful evidence for his theory when, in 1801, he carried out his double-slit experiment. He shone a beam of light through two slits in a piece of card. The slits divided the light into two beams that, when recombined, formed a pattern of alternate light and dark bands on a screen. Only Young's wave theory could explain this result. He reasoned that the pattern was produced by "interference" between the waves in the two beams. Many scientists accepted his explanation, and the wave theory of light gained wide recognition. Today, scientists believe it is impossible to really pin down the nature of light; sometimes it behaves as waves and sometimes as particles. For most practical purposes, though, light can be regarded as waves, exactly as Young suggested.

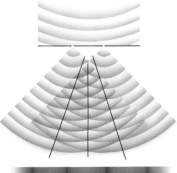

Laser art
Films and rock concerts use lasers to produce striking special effects.

■ Interference

Light waves spread out from their source in ever-increasing circles, like ripples on a pond. Where two waves meet, they combine in some places and cancel out in others. This process, called interference, produced the light and dark bands Young saw in his double-slit experiment. The waves from each slit traveled the same distance, so their crests and troughs were in step. Where two troughs met, they reinforced each other and produced a bright band. Where a crest met a trough, the waves canceled each other out and produced a dark band.

On the crest of a wave (left)
An interference pattern produced by red laser light. The laser beam passes through a diffraction grating, which is a surface containing hundreds of microscopic parallel slits. The slits divide the light into several beams, which spread out, or diffract. These separate beams interfere where they meet.

The double-slit experiment
In Young's experiment, a single light source was split by a double slit. The two beams interfered, producing alternate dark and light bands.

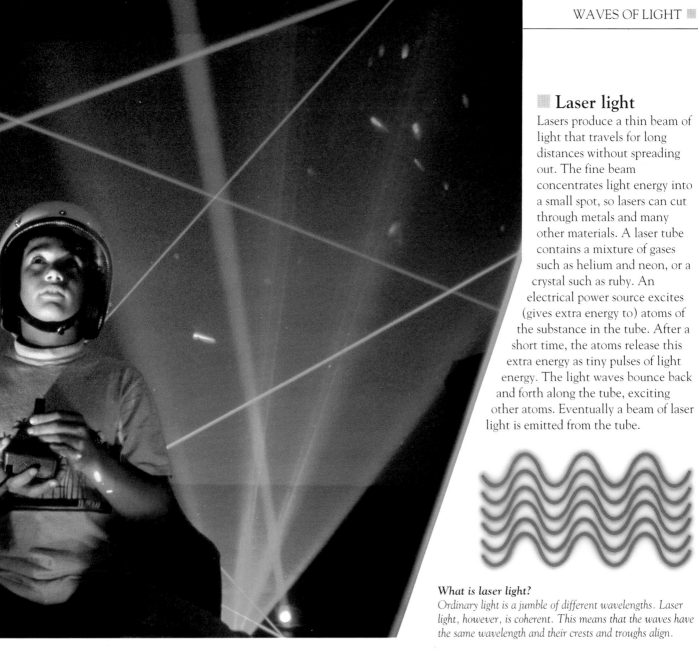

■ Laser light

Lasers produce a thin beam of light that travels for long distances without spreading out. The fine beam concentrates light energy into a small spot, so lasers can cut through metals and many other materials. A laser tube contains a mixture of gases such as helium and neon, or a crystal such as ruby. An electrical power source excites (gives extra energy to) atoms of the substance in the tube. After a short time, the atoms release this extra energy as tiny pulses of light energy. The light waves bounce back and forth along the tube, exciting other atoms. Eventually a beam of laser light is emitted from the tube.

What is laser light?
Ordinary light is a jumble of different wavelengths. Laser light, however, is coherent. This means that the waves have the same wavelength and their crests and troughs align.

■ Polarized light

Light from an ordinary source, such as a bulb, consists of waves that vibrate in many different directions. Special filters produce polarized light, in which the waves vibrate in only one direction. Polarized light can reveal mechanical stress in plastics. The molecules in the plastic split light into two rays that are polarized in slightly different directions. The waves interfere with each other and produce a pattern of colored bands. Engineers study plastic models of aircraft and buildings, using polarized light in order to locate the stresses within them.

Under stress (right)
When a plastic ruler and dish are placed between two polarizing filters, colored interference bands reveal the stress contours in the objects.

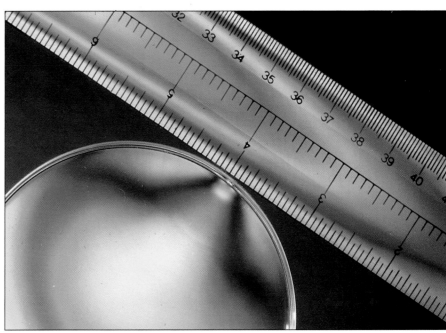

SOUND

WE LIVE IN A WORLD FILLED WITH SOUND. Noise levels near the runway of a busy airport can be high enough to cause physical damage to your ears. But even in the quietest moments, some sounds inevitably reach our ears, whether it is the rustling of leaves or just the sound of your own breathing.

People guessed that sound was vibrations in the air as long ago as the days of Ancient Greece, 2,500 years ago. The famous Greek philosopher Aristotle (384–322 B.C.) believed that light and sound traveled through the air like waves in the sea. If this is so, he deduced, neither light nor sound can pass through a vacuum, where there is no air to transmit them.

However, it was almost 2,000 years before anyone was able to create a vacuum to prove Aristotle right or wrong. Indeed, Aristotle was convinced that a vacuum could never be made. In the early 17th century, though, scientists tried to evacuate air from glass jars to discover whether a bell would ring inside. But they could never remove all air from the jar.

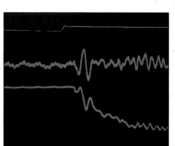

Sound travels through the air in waves, squeezing and stretching the air as it goes. The crests where air is squeezed (compression) and the troughs where air is stretched (rarefaction) can be seen clearly on an oscilloscope trace. The biggest peaks are the loudest sounds.

Silent bell

Finally, in the 1640s, Evangelista Toricelli (1608–47) succeeded in creating a vacuum (p. 117). It was at once obvious that Aristotle had been wrong about light. Light shone through the vacuum without any hindrance at all. But if Aristotle was wrong about light, he was right about sound. In 1654, Otto von Guericke (1602–86), mayor of Magdeburg in Germany, invented an air pump and successfully conducted the bell-in-a-jar experiment. Once the pump had sucked all the air from the jar, Guericke could see the bell ringing noiselessly in the vacuum.

The great Irish scientist Robert Boyle (1627–91) confirmed Guericke's results with a famous experiment. Using an improved version of Guericke's air pump made for him by the brilliantly inventive physicist Robert Hooke (1635–1703), Boyle slowly pumped air from a glass globe containing a loudly ticking watch.

As Boyle pumped out the air, he listened to the sound of the watch becoming steadily fainter. Soon he could see the watch hands turning inside the globe, but could hear absolutely nothing. He then let air back into the globe, and heard the sound of the ticking return. So it was clear that sound needs air to be transmitted.

When a large gong is struck, the vibrations that create sound waves in the air can be seen clearly.

In fact, sound travels even better through solids and liquids than it does through air, but most sounds reach us through the air. What Boyle's test showed is that sound is carried by the material it travels through. Just how this happens is described on p. 104.

Speed of sound

Over 1,900 years ago, the Roman science writer Gaius Pliny (23–79) decided that light must travel faster than sound, because he could see a flash of lighting in a thunderstorm sometime before he heard the crash of thunder. You can see this is true for yourself. Next time there is a thunderstorm, try working out how long thunderclaps take to reach you.

The shape of a bell amplifies air movements started by the impact of the clapper on metal.

Light travels so fast that you see the lightning flash almost as it happens. But because sound travels much more slowly, it takes a little while to arrive. Since we now know how fast sound travels, you can work out how far away the storm is by timing the interval between the lightning and the clap of thunder. Sound travels at 1130 ft (344 m) per second, so the storm is about one mile away for every five seconds you count (one km for every three seconds).

The first serious attempts to establish the speed of sound began in the early 17th century. A Francescan friar named Marin Mersenne (1588–1648) tried two methods. One involved shouting at a wall a known distance away and working out how long afterward he heard the echo. The other was to get a colleague to fire a gun a known distance away.

Bats navigate in pitch darkness with the aid of echoes of ultrasonic squeaks.

Brother Mersenne would then work out how long after he first saw the flash he could hear the bang. There was at that time no way of timing seconds accurately, but the friar made rough estimates by counting his heartbeats. This way, he arrived at a speed for sound of around 1480 ft (450 m) per second.

A century later, in 1738, members of the Paris Academy fired two cannons 18 km (11.3 miles) apart and measured the time between the flashes and the sound arriving at the other cannon. From this, they deduced that sound travels at about 1100 ft (336 m) per second. Since they recorded a temperature of 0°C (32°F) on the day of the experiment, this accords well with most modern estimates.

Varying speeds

Old phonographs recorded sound vibrations as bumps in grooved plastic.

In 1740, an Italian Count, Brianconi, showed that sound travels faster through warm air, and modern scientists estimate that sound travels about 1130 ft (344 m) per second at 68°F (20°C). Through solids and liquids, however, sound moves much faster. In pure water, the speed of sound is almost 1500 m (5000 ft) per second. In solid steel, it reaches nearly 20,000 ft (6000 m) per second. It was once thought that high-pitched sounds must travel faster than lower-pitched sounds. In fact, they travel at exactly the same speed.

■ Ultrasound

During the 20th century, scientists have discovered that there is a whole world of sound that we cannot hear, even with perfect hearing. There is sound pitched higher than even the most sharp-eared human can hear called "ultrasound," and also sound lower than anyone can hear called "infrasound."

But if these sounds are inaudible to us, they can be heard — and made — by a number of other creatures. Bats and dolphins, for instance, can hear ultrasounds with a frequency (pitch, p. 106) more than six times the highest humans can hear. The upper limit of human hearing is about 20,000 Hz (waves per second), but bats and dolphins can make squeaks of 120,000 Hz or more, and hear them too.

One of the special qualities of ultrasound is that it does not spread out nearly so much as ordinary sound, and can be directed almost like a beam of light. This is why ultrasonics are so valuable to bats and dolphins. Ultrasound enables bats to "see" in pitch darkness, and dolphins to find their way underwater.

When flying through the dark, a bat makes constant ultrasonic squeaks. These squeaks echo (bounce back) from any objects in their path, and the sharp ears of the bat pick them up again. From the time it takes for the echo to return and its direction, the bat can build up a sound picture that enables it to fly unerringly through pitch-dark caves.

Sonar equipment on ships and submarines (p. 110) uses ultrasound echoes in much the same way, and all kinds of new uses are now being found for ultrasonic echo location. In industry, for example, ultrasound echoes can detect invisible flaws inside solid metal. The flaw reflects an ultrasonic ping just as an object underwater would, and ultrasound equipment can pick up hairline cracks in nuclear reactors which could be disastrous if left undetected. In medicine, ultrasound echo pictures open a window on the womb of a pregnant woman, allowing doctors to check that the unborn child is developing normally. Unlike X-rays, ultrasound has no known harmful effects, and provides a wonderful moving "live" image.

With equipment using ultrasound echoes to locate things, squeaks tend to be very low-powered. But high-energy ultrasound "guns" can shatter steel girders. Special ultrasound guns called "lithotripters" are already used in medicine to break up kidney stones without surgery. The lithotripter emits a short ultrasonic burst focused on the kidney. The healthy soft tissue of the kidney is unaffected by the sound, because it squeezes and stretches with each wave of sound. But the stones are too rigid to be squeezed, and the high-pitched sound vibrations shake them to pieces.

Supersonic jet planes like Concorde fly faster than the speed of sound. As they accelerate to supersonic speeds, they break the sound barrier, squeezing the air and creating a loud "sonic boom" (p. 107).

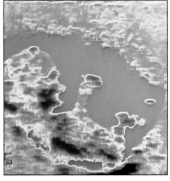

Ultrasound scans give a striking image of a growing fetus inside a woman's womb. Images are often colored artificially by a computer like this one.

What is sound?

EVERY SOUND YOU HEAR is created by something vibrating. Sometimes you can see the vibration; sometimes you cannot. Pluck a taut rubber band, and you can see the band twanging rapidly to and fro as it hums. Clap your hands, or stamp your feet, and you see nothing moving. Vibration is there all the same. But it is not only the band or your hands that move. You hear a sound because air moves too. As the source of a sound vibrates to and fro, it pushes the air molecules around it to and fro. These, in turn, set air molecules around them vibrating — and so the sound is transmitted through the air to your ears in little vibrations or sound waves. Sound waves can travel through solids and liquids as well as air. In fact, sound travels faster in solids and liquids than it does in air, because the molecules are more closely packed. Only in a vacuum is there always complete silence, because sound cannot travel at all without molecules.

Crash, bang, wallop!
When you hit a cymbal, it vibrates rapidly, pushing out a series of vibrations or sound waves through the air.

■ Sound waves

Sound waves are not like waves in the sea. Waves in the sea go up and down, with crests and troughs rippling across the surface. Such waves are called transverse waves. Sound waves, however, are longitudinal. This means that they move by alternately squeezing and stretching, like the spring below, not by going up and down. When a sound is made, the air molecules near the sound are squashed together. They, in turn, jostle up against the molecules next to them, and then are pulled back into place by the molecules behind them. In this way, waves of squeezing and stretching move through the air.

■ How we hear sounds

Without ears, sound would be nothing but vibrations in the air. Ears respond to the thousands of different vibrations in the air and send nerve signals via the auditory nerve to the brain, which interprets them as sounds. Because we have two ears, we can pinpoint fairly accurately just where most sounds are coming from. Any sound reaches one ear just a little before it reaches the other. From the difference between the time the sound arrives at each ear, the brain can work out its origin precisely.

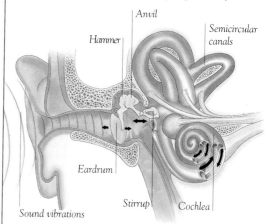

Anvil
Hammer
Semicircular canals
Eardrum
Sound vibrations
Stirrup
Cochlea

Most of the ear is inside the head. The flaps on the outside merely funnel sound down a tube to the eardrum — a skin stretched taut across the tube like the skin of a drum. So thin is the eardrum that even the tiniest sound wave vibrates it to and fro, and as it vibrates, it swings a tiny bone called the hammer against another bone called the anvil. The anvil in turn shakes a third little bone, called the stirrup.

In this way, the vibrations are amplified and transmitted to the inner ear, or cochlea. The cochlea is coiled like the shell of a snail and lined with minute hairs. It is also filled with liquid so that sound travels through it very easily. As the waves of sound pulse through the cochlea, they waggle the hairs. The waggling hairs stimulate tiny nerve cells, which then fire off signals along the auditory nerve to the brain.

Air molecules are squeezed together in some places…

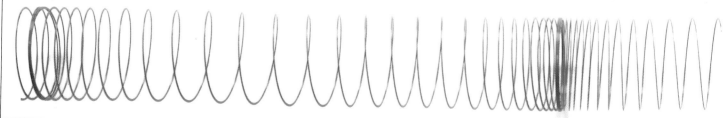

EXPERIMENT
Seeing vibrations

You cannot see sound waves in the air, but you can see their effects. This experiment with sugar dancing on a homemade drum shows clearly that sound is simply vibration.

YOU WILL NEED
- *large cake or cookie tin* ● *sheet of plastic*
- *strong rubber band* ● *baking tray*
- *wooden spoon* ● *brown sugar*

1 Make a drum by stretching a piece of plastic over a large round tin, such as a cake tin.

2 Stretch the rubber band around the tin to hold the plastic taut.

3 Sprinkle a teaspoon of brown sugar on top of the plastic drumskin.

4 Hold a baking tray above your drum, and tap it smartly with a wooden spoon.

You will see the sugar dancing up and down on the drumskin.

■ How it works
When you bang the baking tray, the metal continues to vibrate for a fraction of a second afterward. As it vibrates, the air around is vibrated too. These little vibrations in the air (sound waves) quickly work their way out through the air in all directions. When they hit the drumskin, they set that vibrating too, so the sugar dances up and down on the drumskin. The sound waves that reach your ear make you hear the bang.

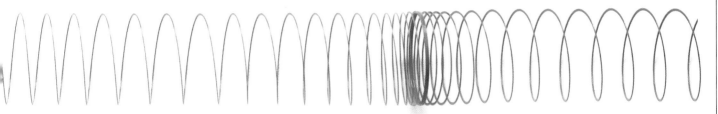

…and stretched out in others…   *… and squeezed together…*   *… and so the sound moves along in waves.*

Pitch and tone

WHEN A BASS DRUM BOOMS, you hear a low-pitched sound; when a seagull screeches, you hear a high-pitched sound. What makes them different is, essentially, the "frequency" of their sound waves. If vibrations follow each other only slowly, the sound tends to be low in pitch; if they follow each other in rapid succession, the sound is high-pitched.

Usually, frequency is measured in hertz (Hz) — that is, cycles (waves) per second. Our ears can hear sounds down to about 20 Hz and up to around 20 kHz (20,000 Hz). Sound any higher than 20 kHz is called "ultrasound," and even though we cannot hear it, it can be detected by bats and dolphins. Ultrasound echoes can also be used to build pictures on a screen — a technique used to locate cracks in buildings and detect abnormalities in a fetus in the womb.

Very few sounds contain notes of one pitch alone. Most have a basic or "fundamental" pitch and a whole series of less important "overtones." It is these overtones that enable us to distinguish one sound from another.

Echo-location
A bat hunting for food sends out a high-pitched sound, which is reflected back from its prey as an echo. The echo enables the bat to pinpoint its prey.

■ Frequency and waves

High-frequency notes do not travel any faster than low-frequency notes; both travel at the speed of sound. Although high-frequency waves come in quick succession, they are much shorter than low-frequency notes. Think of an adult and child walking together. If they walk at the same speed, the child has to take more steps because his or her steps are shorter.

Pitch, though, does not depend only on frequency. Very low notes sound lower if they are loud, for example, while very high-frequency notes sound even higher if they are loud.

Bottle band
Arrange several same size bottles in a row, and fill them with decreasing amounts of water. Blow across each bottle to start the air columns inside vibrating. Listen to the sound each one makes. Which bottle emits the highest note, and which the lowest? The shorter the air column, the higher the pitch.

EXPERIMENT
The Doppler Effect

If you listen carefully to the sound of a police car speeding past you with its siren blaring or a train roaring past another train, you may hear something strange happening to the pitch. Instead of staying constant, the sound seems to get higher as the car approaches, then lower as it goes past. In fact, there is no change in pitch; it just seems so because sound waves reach you faster as the car gets nearer. This is known as the Doppler Effect, after Austrian physicist Christian Doppler (1803–53) who first described it 150 years ago. You can demonstrate this effect very simply yourself.

YOU **W**ILL **N**EED
● *a friend on a bicycle* ● *a whistle*

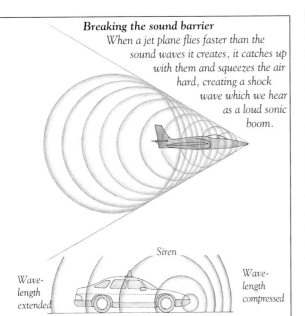

Breaking the sound barrier
When a jet plane flies faster than the sound waves it creates, it catches up with them and squeezes the air hard, creating a shock wave which we hear as a loud sonic boom.

Siren

Wave-length extended

Wave-length compressed

Up and down
As a police car comes toward you, it squeezes the sound waves from the siren, making them sound higher than they really are. As the car moves away again, the sound waves stretch out behind, making the siren sound lower.

1 GET A FRIEND to ride past you slowly on a bicycle, blowing a whistle.

2 LISTEN CAREFULLY for any change in the pitch of the whistle.

3 NOW ASK your friend to ride past very quickly, still blowing the whistle. This time you should hear the whistle note rising and falling.

■ Tone
Even when they play a note of exactly the same pitch, a flute and a violin sound very different. This is because every musical instrument produces its own distinctive range of overtones as well as the fundamental note. The pattern of sound waves from a musical instrument can be represented as wavy lines, with peaks and troughs showing the change in air pressure as the sound reaches your ear. Each instrument has its own distinctive wave pattern.

Flute
The flute produces a much purer and mellower sound than the violin, with only a few very regular overtones.

Gong
Hitting a gong makes it vibrate in an irregular pattern, creating a crashing sound with a jagged, messy wave pattern. We hear it as a noise with no clear pitch.

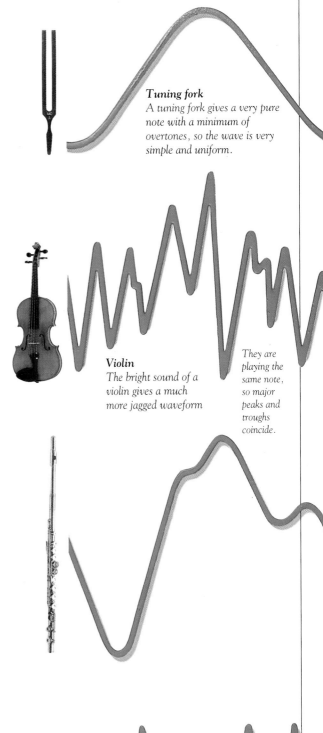

Tuning fork
A tuning fork gives a very pure note with a minimum of overtones, so the wave is very simple and uniform.

Violin
The bright sound of a violin gives a much more jagged waveform

They are playing the same note, so major peaks and troughs coincide.

Resonance

EVERY OBJECT has its own natural frequency. In other words, if allowed to vibrate freely, it always tends to vibrate at the same rate. If you strike a bell, for instance, it will always give the same note. Similarly, the pendulum in an old clock always tends to swing to and fro at the same rate. You can make objects vibrate faster or more slowly than their natural frequency by jogging them at appropriate intervals. This is called "forced vibration." But when an object is jogged at just the same rate as its natural frequency, it vibrates in sympathy and the vibrations become much stronger. This is called "resonance." It is like pushing a swing. Time your pushes too early or too late, and it is hard work to keep the swing going; time them well, and it all becomes effortless. This is why trains often trundle slowly across long bridges — in case the vibration should match the bridge's natural frequency and shake it to pieces. It is also why an opera singer can shatter a wine glass by singing a note at the natural frequency of the glass.

Ringing out
Every bell has its own natural frequency and so always rings at a particular pitch.

■ Measuring sound

The loudness of a sound depends on just how much energy there is in the sound waves. Big, energetic waves move your eardrums a long way and sound loud; smaller waves move them much less and sound quiet. Sound energy or "intensity" is sometimes measured in "bels" (after the Scottish-American inventor Alexander Graham Bell, 1847–1922), but more usually in decibels (0.1 bel). The decibel scale is logarithmic. This means that a 2-decibel (dB) sound is not twice but ten times as intense as a 1-dB sound, and a 20-dB sound is 100 times as intense as a 10-dB sound. Sounds over about 120 dB can cause intense pain and deafness.

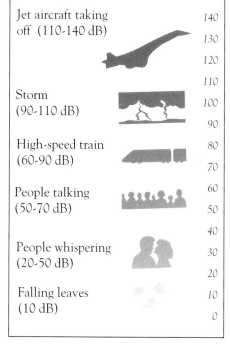

Jet aircraft taking off (110-140 dB)	140
	130
	120
	110
Storm (90-110 dB)	100
	90
High-speed train (60-90 dB)	80
	70
People talking (50-70 dB)	60
	50
	40
People whispering (20-50 dB)	30
	20
Falling leaves (10 dB)	10
	0

■ Blowing a flute

When a musician blows a flute, or any wind instrument, it makes the air inside resonate — that is, vibrate at its natural frequency. With woodwinds the player blows over a reed to set the air in motion; with brass, there is a metal mouthpiece. Just how fast the air vibrates — and what note sounds — depends on the length of the column of air inside the pipe. If it is short, the air vibrates quickly, giving a high note. If it is long, it vibrates slowly, giving a low note. The player can alter the note by fingering holes or closing keys to change the length of the column. Vibration is at a peak at each end of the column. These peaks are called "antinodes." Vibration drops to zero in the middle (the "node").

Peak vibration of air (antinode)

Zero vibration of air (node)

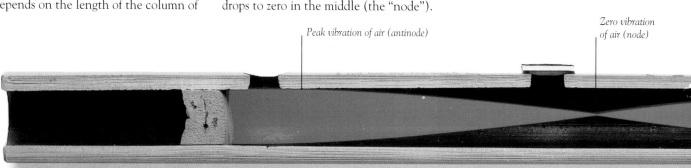

EXPERIMENT
Pendulum Power

You can set up your own experiment to illustrate the power of resonance.

YOU WILL NEED
● *string* ● *three identical weights*

1 Make three pendulums by attaching different lengths of string to each weight, and tie them at equal distances along string pulled taut between two secure points. Swing each of the pendulums in turn, and watch what happens.

2 Repeat the exercise — only this time hang two pendulums from identical lengths of string. As either of the two matched pendulums swings, its partner will begin to swing too — because it resonates.

3 Repeat the exercise with the pendulums all at equal lengths. Now all three will resonate and swing together.

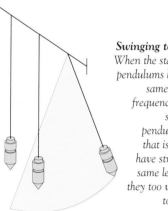

Swinging together
When the stationary pendulums have the same natural frequency as the swinging pendulum — that is, if they have strings the same length — they too will start to swing.

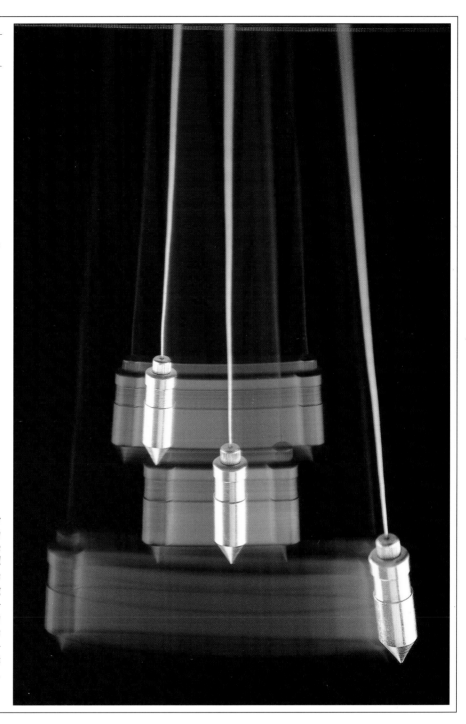

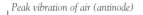
Peak vibration of air (antinode)

Echoes and acoustics

WHEN YOU SHOUT in a large, empty hall or in a tunnel, you can sometimes hear a noise still ringing out a moment or two later. This is simply the sound of your voice bouncing back from the walls, for just as light can be reflected, so sound

Round echo
In the whispering gallery of St Paul's Cathedral, London, you can hear a whisper on the far side because the dome focuses all the echoes together.

echoes. A small room echoes just as much as a large hall, but you can hear an echo only if it comes back at least 0.1 second after the original sound. As sound travels about 33 m (36 yards) in 0.1 second, you only hear echoes from surfaces at least 17 m (18 yards) away. Smooth, hard surfaces give the best echo because they break up the sound waves least.

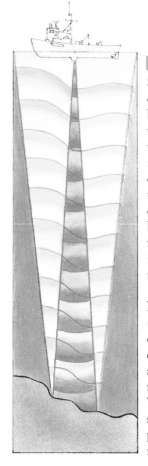

■ Echo sounding

Ships can use echoes to chart the depth of oceans, because sound travels very well underwater. Sound waves, produced by an instrument called an oscillator, are bounced off the ocean bed. The depth of the water is estimated by an instrument called a fathometer from the time it takes for the echoes to return to the ship.

Similarly, Sound Navigation and Ranging (Sonar) equipment uses echoes underwater to detect the location of anything from shoals of fish to enemy submarines. The sonar sends out a stream of high-frequency pings and analyzes the returning echoes.

■ Acoustics

In a concert hall, the ultimate quality of the sound reaching the audience depends on its acoustics — that is, the way sound echoes around the hall. So before any new hall is built, the acoustics of the design are usually tested with tanks of water made in exactly the same shape. Water waves behave in many respects like sound waves, and the ripples in the tank give a good idea of how sound will echo around the hall.

The idea is not to eliminate echoes altogether, but to use them to create the best possible sound, and to ensure that everyone in the audience can hear properly. Without echoes, the music would tend to sound rather flat and lifeless. So even in the best-designed concert halls, the music can often be heard fading away fractionally after the orchestra stops playing. This is called the reverberation time. Concert halls are generally designed to give a reverberation time of two seconds or so; in big churches and cathedrals, it may be anything up to eight seconds, giving a mellower sound.

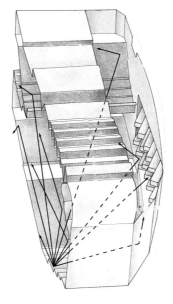

Concert hall acoustics (above)
This cutaway concert hall has a suspended ceiling and angled boxes to help reflect sound from the stage to the audience with very little distortion. Designers of concert halls limit the use of soft furnishings, as these tend to absorb sound, and avoid too many hard, flat surfaces that cause reverberation.

EXPERIMENT
Bouncing sound

YOU WILL NEED
● *two stiff cardboard tubes (poster/art tubes)* ● *ticking clock or watch* ● *large smooth piece of card (15¹/₂ in / 40 cm square) and pieces of fabric and foam*

This experiment provides a convincing demonstration of how smooth, hard surfaces reflect sound in the same way that a mirror reflects light, while fabric and foam reflect sound poorly.

1 Hold the card upright on a large table, and place the tubes at an angle to the card. Leave a gap of about 2¹/₂ in (6 cm) between the card and the ends of the tubes.

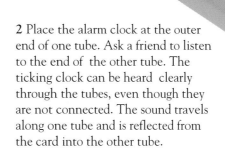

2 Place the alarm clock at the outer end of one tube. Ask a friend to listen to the end of the other tube. The ticking clock can be heard clearly through the tubes, even though they are not connected. The sound travels along one tube and is reflected from the card into the other tube.

The curved strip represents the back wall of a stage.

The ripples represent sound waves, radiating in concentric circles from the source.

Where the sound waves meet the curved strip, they are reflected in a straight line, projected directly away from the back wall.

EXPERIMENT
Making a ripple tank

Sound waves normally radiate from their source in circles. But a curved surface of precisely the right shape will reflect them in straight lines, so that they reach everywhere along the line at the same time. In concert halls, the back of the stage is sometimes curved to create this effect. You can see how it works with ripples in water.

YOU WILL NEED
● *large deep baking tray* ● *long, thin strip of metal or Formica* ● *pipette (eye dropper)* ● *black ink* ● *water*

1 Fill the baking tray with water mixed with black ink.
2 Bend the strip into a curve to represent the back of the stage.
3 Let a series of drops of water from the dropper fall as evenly and accurately as possible in the center. Watch the ripples reflected from the curve.

The sound of music

ALL SOUNDS are created by something vibrating, and the sounds from musical instruments are no exception. But the sound of music has a very special quality that makes it different from plain noise. The vibrations of air in a flute occur at regular intervals, while those from a car engine do not. It is the regularity of the vibrations that distinguishes music from noise. Musical instruments produce these regular vibrations many different ways. Stringed instruments, for instance, have taut strings that vibrate to and fro when plucked or bowed. In wind instruments, the vibrations are created when the player blows air across a hole or a reed, or into a specially shaped mouthpiece.

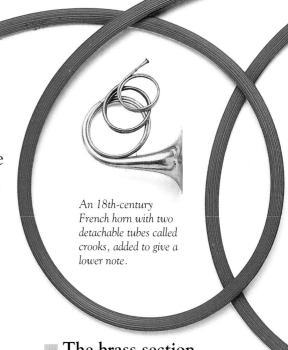

An 18th-century French horn with two detachable tubes called crooks, added to give a lower note.

Complete orchestra
Simple instruments from all sections of the orchestra can be made very easily at home.

■ The brass section

In brass instruments, such as trumpet and horn, sound is created by the vibration of the player's lips against the mouthpiece. As the lips vibrate, pulses of air are funneled through the mouthpiece to the air column inside. With horns and bugles, different notes are made by changing the shape of the mouth and blowing hard or soft.

■ The string section

Simple forms of stringed instruments have been used for thousands of years, and the basic science of vibrating strings was first worked out by Pythagoras around 500 B.C. He realized that the sound produced depended on the size of the string and how tightly it is stretched. Different sounds are produced by changing the length of the strings and pressing them down at different points. This is called "stopping" the string. The closer in a string is pressed, the shorter the part that is vibrated, and the higher the note.

Making a guitar (right)
Make a circular hole in a hollow box, such as a shoebox. Fold some cardboard to make the "bridge," and fix it above the hole as shown. Insert six paper fasteners at each end of the box, and stretch six strong rubber bands across the box and bridge, winding them around the paper fasteners. Pluck the bands to make a sound. Stretch them tighter, and a higher sound is produced.

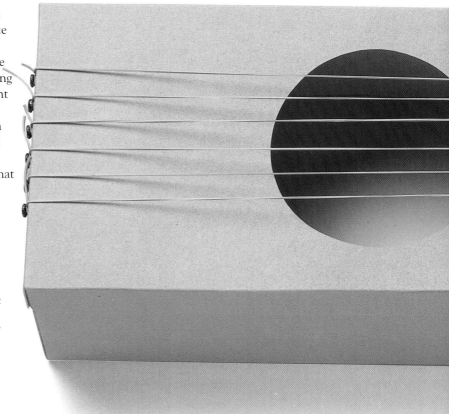

Making a horn

You can make a very simple horn by attaching a funnel to the end of a piece of hose piping. The difficult part is making a good sound. Try pursing your lips tightly together, then blow through them so that they vibrate quickly, making a sound like a trumpet. Then curl the tubing around your shoulder and hold the funnel end up. Now, keeping your lips tightly pursed, press them up against the end of the tube and blow hard. With practice, you should be able to sound a rounded note.

Making panpipes

Take seven drinking straws (the wider the better), and line them up ½ in (1.3 cm) apart. Cover them in the middle with adhesive tape. Tune them by cutting each one slightly shorter than the one before, at even intervals. Blow gently across the top of each straw to produce a note.

The wind section

There are many kinds of wind instruments, from ancient panpipes to modern saxophones. But they all create sound by vibrating a column of air inside a tube. Different notes are made by changing the length of the air column. Panpipes and organs have a different tube for each note. With brass and woodwind instruments, the player can play higher notes just by blowing harder, making the air column split into halves or thirds. Notes in between can be made by blocking off or opening holes in the tube to alter slightly the length of the air column inside. In some instruments, like the flute, both ends of the tube are open. In others, such as organs, the tube is closed at one end. A "stopped" tube gives a note that is an octave (eight notes, or a full scale) lower than an open tube of the same length.

Panpipes

Panpipes consist of graduated pipes, each producing one note. According to Greek legend, the god Pan made these pipes when the nymph that he loved was changed into a reed.

Shaker

Maracas can be made from empty yogurt containers filled with seeds or beads. Seal the ends with colored tape.

Gong

A gong is a large metal plate, suspended and struck at the center. The sound vibrates from the center to the edges. Improvise a gong with a metal bowl and a wooden spoon.

Tambourine

A tambourine has small metal plates (jingles) that rattle when the instrument is shaken. A "jingling stick" can be made by fixing some bottletops to a wooden stick or handle. This instrument is often used in traditional folk dances.

The percussion section

Most early instruments were percussion instruments, the kind that make a sound as a result of being hit or shaken. What they all usually have in common is a hollow space within which the sound can resonate (see p. 108), even if it is just the inside of a yogurt container. The first was probably a hollow log hit by a stick. Later, animal skins were stretched over objects and set vibrating by hand. You can add many types of percussion instruments to your orchestra, using a variety of materials — wood, metal, pebbles in a jar — that make a full sound.

Air
■ AND ■
water

AIR AND WATER are among the most important substances on Earth. Water covers about 70 per cent of the Earth's surface and makes up a similar proportion of our bodies. Air is the almost invisible mixture of gases in which we move, live, and breathe.

The spiraling storm clouds of a hurricane in the North Atlantic bring air and water together in a combination of awesome power and energy.

AIR

AIR IS SO TRANSPARENT that we can easily forget about it. Yet it is real and substantial. Without air, life on Earth would be impossible. Air not only provides the oxygen that humans and all other animals need to breathe; it is also a major part of our environment. The atmosphere of air surrounding our planet also protects us from the harmful effects of cosmic rays and meteors, and, like a fluffy comforter, keeps the Earth's surface warm.

The reddest, most colorful sunsets and sunrises occur because there are particles in the air that scatter away other colors of light from the sun. Some of the most beautiful recent sunsets in the northern hemisphere occurred after vast clouds of dust thrown up by the eruption of Mount St. Helens in Washington were spread around the world by winds high in the atmosphere. For the same reason, industrial areas often have stunning sunsets.

Air was one of the four elements into which the ancient Greek philosopher Aristotle (384–322 B.C.) divided the world nearly 2500 years ago (p. 12). But right up until the 17th century, no one knew very much about it. Most thinkers assumed it was a simple, uniform substance — not the complicated mixture of different gases and particles we now know it to be.

It was perhaps the Belgian alchemist Jan Baptista van Helmont (1579–1644) who first began to suspect the truth. Indeed, Helmont is said to have invented the word "gas" in about 1620 — taking it from the Greek word *chaos*. The Greeks thought chaos was the strange substance out of which the universe was originally created.

■ Pure air?

Like Aristotle, and like other alchemists, Helmont believed air was one of the basic elements. Yet he saw that the vapor given off by fermenting fruit juice is a chemical in its own right, not just air. In fact, the vapor is the gas carbon dioxide, although it was another 100 years before the Scottish chemist Joseph Black (1728–99) finally proved this.

Half a century after Helmont, about 1670, an English doctor named John Mayow (1641–79)

conducted a now famous experiment to find out more about burning and breathing — both of which need air. He burned a candle in air trapped in a jar turned upside down in water. As the candle burned down, so the water in the jar rose, showing that burning actually consumed air. He found a similar rise in the water level when he tried the experiment with a live — breathing — mouse in the jar instead of a burning candle.

■ Burning

Rightly, Mayow concluded that air contained a substance that was needed for burning and breathing, which he called the "nitro-aerial spirit." Once this was used up, candles would not burn and mice (and humans) could not breathe.

Sadly, Mayow's ideas were forgotten, for around the same time as he was experimenting with candles and mice, the German chemist Johann Becher came up with another idea to explain combustion (burning). Becher suggested that everything burnable contained a special substance that was released as it burned. He called this substance *terra pinguis*, or "combustible earth." Wood, Becher

The pressure of air is enough to crumple a metal can (p. 123).

thought, is made of terra pinguis and ash — so when it burns, the terra pinguis is released, leaving just ash.

Forty years later, the German chemist Georg Stahl suggested that terra pinguis should be called *phlogiston* instead, from the Greek word for "burned." The name stuck. Indeed, for almost a century, most scientists were convinced that the phlogiston theory fully explained burning and many other chemical reactions, such as rusting.

In Mayow's experiment (p. 119), water rose in the jar as the candle burned, showing some air was used.

■ Oxygen

It was not until the 1770s that people really began to doubt the truth of the phlogiston theory. On August 1, 1774, the British minister and amateur chemist Joseph Priestley made another famous experiment, at Bowood in Wiltshire, southern England. Priestley heated mercuric oxide, and found it gave off a colorless gas that made a candle burn dazzlingly bright.

A few months later, he wrote that "two mice and myself have had the privilege of breathing it." Priestley had discovered oxygen.

Priestley did not fully appreciate the significance of his discovery, and still believed in

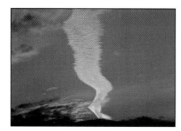

Water vapor is present in varying amounts throughout the lowest layers of the atmosphere. When the air is cold or damp enough, this condenses to drops of water, forming clouds. Most are natural, but some, like this contrail, are made by jet planes.

phlogiston. But in October, he told the brilliant French chemist Antoine Lavoisier about oxygen. Lavoisier immediately repeated and improved on Priestley's experiment and demolished the phlogiston idea once and for all — although some scientists took a great deal of convincing.

Lavoisier decided that air was made up of at least two gases. One fifth of air was the same gas Priestley had found, which Lavoisier named oxygen. The remainder was the gas later known as nitrogen, which Lavoisier called *azote*, from the Greek for "no life." Oxygen, Lavoisier showed, was the substance needed for burning, not phlogiston. And it was oxygen from the air that was consumed by burning, not phlogiston in the burned substance.

Combustion

Now we know that Lavoisier was essentially right. When something burns in air, it combines with oxygen in the air — a process called "oxidation." In fact, the oxygen need not come from the air; it can be chemically combined, for instance, within an "oxide." Combustion can be very rapid, like when a bomb goes off, producing huge flames. But it can also be very slow, producing no flames at all. Breathing is a form of slow combustion. When we breathe, we take in oxygen from the air. This combines chemically with glucose (sugar) in the body to create energy.

For more than a century after Lavoisier's experiments, most scientists believed that air was a mixture of oxygen and nitrogen alone. Moreover, French chemist Henri Regnault

The Sun's heat keeps the air surrounding the Earth forever on the move, creating both gentle breezes and ferocious hurricanes. This is a homemade windvane and meter (pp. 126-127).

(1810–78) showed that air was the same all around the world. Then, in the 1890s, the Scottish chemist William Ramsay (1852–1916) discovered that air also contained tiny traces of "inert" gases (gases that are very slow to react with other chemicals) — first argon, then helium, then krypton, xenon, neon, and radon.

Air pollution

Scientists now know that pure air contains both oxygen and nitrogen, plus all the inert gases. But air is very rarely pure. Mixed in with these gases, there are always traces of other substances or pollutants, varying in concentration around the globe. Some of these are gases, such as carbon dioxide, which plays an important part in the lives of both plants and animals (pp. 119, 121). Some are tiny particles of dust, small and light enough to be held up by the air, like the dust thrown high into the air by erupting volcanoes.

Increasingly, the air is becoming polluted by waste from human activities, especially cars and heavy industry. Cars and trucks, for example, not only push out solid particles of soot into the air; they also emit poisonous gases such as nitrogen dioxide and carbon monoxide. In the short term, air pollution can cause illness and foggy weather. In the long term, many experts fear, the consequences may be disastrous unless we act rapidly to clean up our air (pp. 120-121).

You can measure air temperature with this simple thermometer (p. 125).

Vacuums and pressure

For thousands of years, scientists thought there could be no such thing as a vacuum — a space with nothing in it whatsoever, not even a gas. Aristotle insisted that a vacuum could not exist anywhere in the universe. But in 1640, a little after Helmont discovered carbon dioxide, a young Italian named Evangelista Torricelli (1608–47) created a vacuum very simply.

Torricelli was a pupil of the famous scientist Galileo Galilei (p. 66). Torricelli filled a long tube completely with water, put a stopper in the top, and held it upright in a bowl of water. At once the water level dropped in the tube, leaving a space between the water and the stopper. This, concluded Torricelli, could only be a vacuum.

But why did the water fall only so far down the tube? Torricelli believed that it was held up by the weight of the 50 miles or so (80 km) of air above pressing down on the water in the dish. So Torricelli had not only shown the reality of vacuums, but also suggested that air itself was real enough to exert pressure. He was wrong, however, about air pressure being simply a result of the weight of air. As Robert Boyle showed just 20 years later in 1662, air pressure is related to the *density* of the air (p. 25). Boyle's historic discovery laid the basis for discoveries about the nature of matter which led to the atomic theory (p. 13).

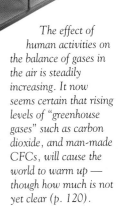

The effect of human activities on the balance of gases in the air is steadily increasing. It now seems certain that rising levels of "greenhouse gases" such as carbon dioxide, and man-made CFCs, will cause the world to warm up — though how much is not yet clear (p. 120).

In the air

ENVELOPING THE WORLD is a thick blanket of gases called the atmosphere. Without this blanket to protect us, we would be roasted by the Sun during the day, then frozen at night as all its heat escaped into space. The outer reaches of the atmosphere are many hundreds of miles above us. But most of the gases are squeezed into the lowest 9 miles (15 km) or so, called the troposphere. Only the troposphere contains water vapor and enough air for us to breathe and survive. Even in the troposphere, air gets very thin high up — which is why climbers on Mount Everest need breathing apparatus.

Airless Moon
Because it has no atmosphere, the Moon gets very hot and very cold.

Satellite

Space shuttle

Auroras

Meteors

Weather balloon

Helicopter Airplane Concorde

Layers of atmosphere
The Earth's atmosphere has five layers.

Exosphere
above 310 miles

Ionosphere
up to 310 miles
(500 km)

Mesosphere (to 51 miles)

Stratosphere (to 31 miles)

Troposphere (to 9 miles)

■ What is the atmosphere made of?

The gases in the atmosphere came originally from volcanoes active when the Earth was young, but the mixture has been slowly modified by living creatures over billions of years. More than three-quarters of the atmosphere (78 per cent) is nitrogen, and most of the rest (21 per cent) is oxygen, which all animals need to breathe. Less than 1 per cent is carbon dioxide, water vapor, and traces of other gases, such as neon, helium, ozone, and krypton. But the tiny quantities of carbon dioxide and water vapor are vital to life on Earth. Both help keep the planet warm by preventing heat from escaping into space. Water vapor also condenses out of the air to form clouds, and so gives the rain on which life is so dependent. Without water vapor, there would be no weather (see p. 124).

Gases in the air
This bunch of balloons shows the proportions of different gases in the atmosphere: blue balloons are nitrogen, red oxygen, and the single white balloon all the other gases.

EXPERIMENT
Put out a fire with carbon dioxide

Adult help is advised for this experiment

Each of the gases in the atmosphere has its own particular properties. Although only present in small quantities, carbon dioxide attracts attention because of the role it plays in the greenhouse effect (p. 120). But carbon dioxide also produces the bubbles in soft drinks and mist effects on stage and television as "dry ice" (p. 16). This experiment shows why carbon dioxide can also be used as a foam to help fight fires.

YOU WILL NEED
● *bicarbonate of soda* ● *vinegar* ● *glass bottle, such as small soft-drink bottle* ● *candle* ● *saucer* ● *modeling clay*

1 FIX THE CANDLE to the saucer with modeling clay. Put the candle on a table, but make sure that it is not near anything that could catch fire. Light the candle. Next put one tablespoonful of bicarbonate of soda into the bottle. You can use a small piece of paper as a funnel. Then pour about three tablespoonfuls of vinegar into the bottle. When the vinegar touches the soda, they will react to form carbon dioxide.

2 THE CARBON DIOXIDE will not float out because it is heavier than air. Now place your thumb over the open end of the bottle, and tilt the bottle over the candle. Be careful not to burn your thumb. Take your thumb away, and slowly "pour" the gas over the flame, taking care not to spill any liquid. The invisible fumes of carbon dioxide from the bottle will extinguish the flame immediately.

EXPERIMENT
Oxygen in the air

Adult help is advised for this experiment.

In an effort to discover what it is in the air that animals need for breathing and flames need to burn, scientists in the 18th century conducted many experiments with air trapped in a glass turned upside down in water. This classic experiment shows that part of the air is used up during burning.

YOU WILL NEED
● *candle* ● *large glass jar (calibrated if possible)* ● *eggcup* ● *shallow dish of water* ● *matches*

1 FIX THE CANDLE into the eggcup with a little melted wax, and place in the middle of the bowl. Fill the bowl about three-quarters full with water, making sure the candle is well clear of the surface of the water. Light the candle.

2 LEAVE THE CANDLE to burn for a couple of minutes. Place the jar over the candle at a slight angle to expel some of the air, as it lowers the level of the water. Note the level of water in the jar at this stage.

3 AS THE CANDLE BURNS, it uses up the oxygen in the air. The candle wax melts to form the elements carbon and hydrogen, which combine with the remaining oxygen to form water and carbon dioxide. As this oxygen is used up, the water level in the jar rises. When all the oxygen is used up, the flame goes out.

Upsetting the balance

WITHOUT CERTAIN GASES in the atmosphere, the Earth would be very, very cold and life would be impossible. These gases, notably carbon dioxide, trap heat from the Sun and stop it from escaping into space — just as glass traps warmth in a greenhouse, which is why it is called the "greenhouse effect." For millions of years, the greenhouse effect has kept the Earth warm, because the amount of carbon dioxide in the atmosphere has remained stable. Now scientists are worried that man's activities are dramatically boosting levels of carbon dioxide and other greenhouse-effect gases. This could be increasing the greenhouse effect so much that the Earth is heating up.

The airy greenhouse
The atmosphere normally keeps conditions right for life on Earth. It traps just the right amount of heat in the same way that the glass in a greenhouse does. Ultraviolet and visible rays from the Sun penetrate the glass walls and are reradiated inside the greenhouse as infrared rays. These rays are trapped by the glass wall, keeping the greenhouse warm. Open windows allow some heat to escape to the outside.

■ The ozone layer

Another source of worry about the future of the Earth's atmosphere is the depletion of the ozone layer. Ozone (O_3) is actually a form of oxygen (O_2). It is a pale bluish gas that occurs naturally in tiny quantities high in the atmosphere. These small quantities protect us from harmful ultraviolet radiation from the Sun, which can cause skin cancer and stop plants from growing. Scientists have worried for some time about the thinning of the ozone layer. In the mid-1980s, satellite pictures revealed a big hole in the ozone layer that appeared each spring over the Antarctic. The main culprits for destroying the ozone layer are man-made CFC (chlorofluorocarbon) gases. These gases are used in aerosol sprays, for cleaning electronic circuits, and inside refrigerators. CFCs are difficult to break down chemically, but high in the atmosphere, sunlight makes them release chlorine, which combines with oxygen atoms in the ozone, destroying it. This is why the effect is especially bad in spring.

The ozone hole
Satellite map showing the hole in the ozone layer that appears annually over the Antarctic.

■ Global warming

Scientists are convinced that the world is getting warmer because of the increase in greenhouse-effect gases — but they cannot agree by how much. Most predict it will warm between 2 and 4°C by the year 2030 unless we do something drastic to cut down the increase. Some fear that global warming will be much worse.

High temperatures do not necessarily mean better weather. In some places, they will bring drought; in others stormy weather as the extra heat stirs up rain and winds (p. 124). Worst of all, the extra warmth may expand the water in the oceans, bringing floods to places like Bangladesh. If warming continues above 5°C, it may even begin to melt the polar ice caps, drowning coastal cities such as Sydney and New York.

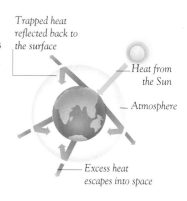

Trapped heat reflected back to the surface

Heat from the Sun

Atmosphere

Excess heat escapes into space

Warm Earth
The Sun's rays warm the surface of the Earth, which radiates heat into the atmosphere. The greenhouse-effect gases in the atmosphere trap some of this heat and radiate it back to the surface.

The stuffy greenhouse
The buildup of greenhouse-effect gases in the atmosphere is like closing the windows of the global greenhouse. It keeps in more heat, so the Earth's temperature increases.

The culprits
Carbon dioxide is the main greenhouse-effect gas, and the burning of fossil fuels such as coal and oil is to blame for much of the increase in the greenhouse effect. Industry and the automobile are both culprits. Particularly unfortunate is the cutting and burning of vast areas of tropical forest to make way for cattle pasture. While alive, trees help absorb carbon dioxide from the air; when they burn, they add more. But carbon dioxide is not the only greenhouse gas. Others are methane from rice fields and refuse dumps and the CFCs that are destroying the ozone layer.

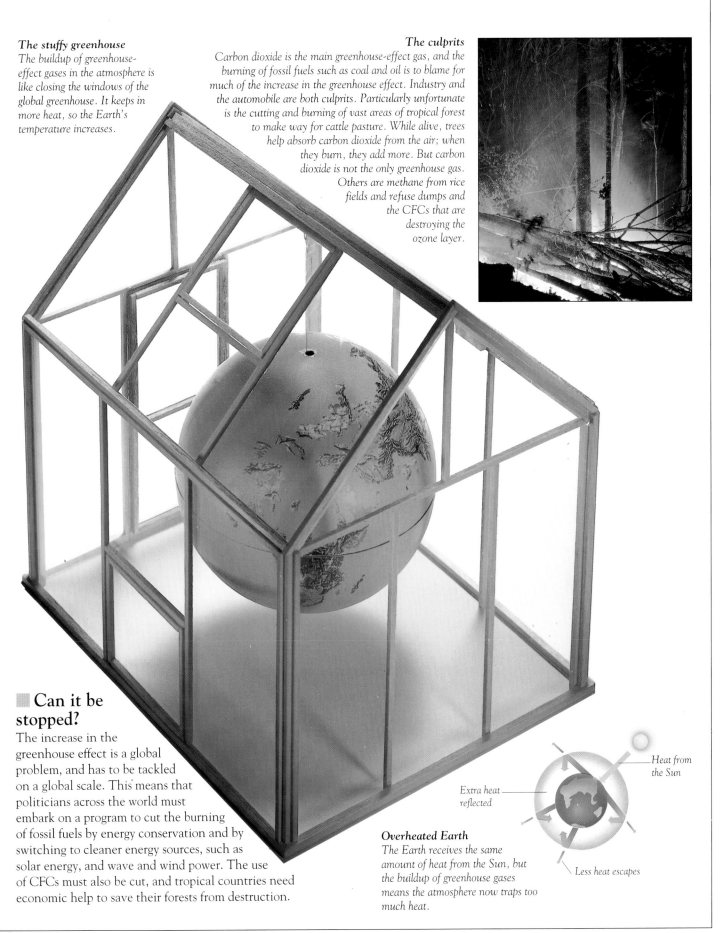

▥ Can it be stopped?

The increase in the greenhouse effect is a global problem, and has to be tackled on a global scale. This means that politicians across the world must embark on a program to cut the burning of fossil fuels by energy conservation and by switching to cleaner energy sources, such as solar energy, and wave and wind power. The use of CFCs must also be cut, and tropical countries need economic help to save their forests from destruction.

Heat from the Sun

Extra heat reflected

Less heat escapes

Overheated Earth
The Earth receives the same amount of heat from the Sun, but the buildup of greenhouse gases means the atmosphere now traps too much heat.

Air pressure

When you swim underwater, you may feel the pressure of water on your body increasing the deeper you go. The air too has pressure, although you cannot feel it. Indeed, the pressure of the atmosphere at ground level is enormous and would crush you if it were not for the fact that the fluids inside your body exert as much pressure as the air pressure outside. Scientists once thought atmospheric pressure was simply the weight of hundreds of kilometers of air pushing down on the Earth. In fact, air

Pushing power
You can prove that air pressure pushes upward as well as down this way. Fill a tumbler to the brim with water, and slide a postcard over the top. Keeping your hand pressed on the card, turn the tumbler upside down over the sink. Take your hand away, and the card will stay in place, held up by air pressure!

pressure pushes in all directions — up, down, and sideways — and it is really constant bombardment by moving air molecules (p. 24). Air pressure gradually decreases as you go higher in the atmosphere — so evenly that aircraft altitude meters work simply by measuring the air pressure. Jet airliners must have pressurized cabins because the pressure drop at high altitudes would make it impossible to breathe — the greater pressure inside your body would prevent it taking in air.

■ DISCOVERY ■
The Magdeburg experiment

IN 1664, THE MAYOR of the German town of Magdeburg, Otto von Guericke, devised an experiment to demonstrate the strength of atmospheric pressure. He used two metal hemispheres, fitted together with an airtight joint, to make a hollow sphere. Once air had been sucked from the inside with a vacuum pump, it took 16 horses to pull the hemispheres apart. This 17th-century engraving shows the Magdeburg experiment demonstrated in the presence of Emperor Ferdinand III.

————— EXPERIMENT —————
Air lift

Rubber suction cups work because of air pressure. When pressed against a flat surface, air is forced out, reducing the pressure inside. Because the pressure outside is now much greater, the suction cup is pushed firmly down. This experiment shows that the pressure is enough to hold a stool in the air.

YOU WILL NEED
● *smooth-topped stool* ● *suction cup*

1 Tie string to the suction cup. Press the cup to the top of the stool.

2 Now try lifting the cup.

DEMONSTRATION
The collapsing can

Air pushes on each and every square inch (6.5 square cm) of your body with a force equal to nearly three bags of sugar – about 14 lbs (6.4 kg). We are normally unaware of the intensity of air pressure because we are supported by equal air pressure on all sides.

Just how strongly air can push is illustrated by this collapsing can demonstration. Air pressure pressing on the outside of a metal can is strong enough to crush the can. Normally, the air inside the can pushes back against this, so the can does not collapse. But if the can is filled with steam, the situation is very different. The steam condenses to water, creating a partial vacuum inside the can. This means that there is very little air to push against the air pressure outside the can, so the can collapses.

How it works
The can retains its shape in normal atmospheric conditions because the air pressure inside the can is the same as the air pressure outside the can.

Creating steam
by boiling water inside the can pushes much of the air out of the can. The pressure of the steam, however, still equals that of the air outside the can.

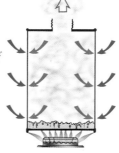

Replacing the cap
allows the steam to cool and condense, creating a partial vacuum. Pressure outside is now greater than that inside, so the can is crushed.

EXPERIMENT
Blowing books away

When you compress (squash) air, the air pushes back. This force can be useful. Hovercrafts, for example, use a cushion of high-pressure air to lift them out of the water. This reduces friction, which means that the vehicles can travel easily and quickly. This experiment shows how compressed air can be used to lift a load.

The harder you blow, the higher the pressure of the compressed air

1 PLACE THE BALLOON on the edge of a table, so that the neck hangs over the edge. Place the books on the balloon. Begin to blow into the balloon.

2 CONTINUE TO BLOW air into the balloon. The pressure of the air inflates the balloon and lifts the book off the table. Can you make the books topple?

Weather

THE LOWEST LAYER of the Earth's atmosphere — the troposphere in which we live — is forever on the move. The Sun warms the Earth's surface, and this heats the air, stirring it into motion. Some places are heated more than others, and there air rises, drawing more air in underneath. Other places are heated less, and there air sinks, pushing air outward. So air circulates continually between warm areas and cold, carrying moisture with it and creating everything that we call weather: wind, rain, mist and fog, snow, hail, and thunder.

The restless air
The constant motion of the air that gives us weather is revealed by swirling clouds in this satellite image.

Forecasting the weather

Modern weather forecasting depends on a multitude of observations all over the globe. Fixed weather stations, buoys and balloons, ships, and airplanes constantly feed back to meteorological centers data about temperature, pressure, wind, rainfall, and so on. The data are then fed into supercomputers to give a complete "synoptic" (seen together) picture of weather throughout the atmosphere. This picture is combined with images from satellites, which show temperature and wind (indicated by the way clouds move). The computers can then give accurate forecasts for up to a week ahead.

Rain
Clouds contain tiny ice particles and water droplets so light that they are held up by the air. But if enough droplets join together, they will fall to the ground as raindrops.

Snow
Inside clouds the air is very cold, and ice crystals often stick together to form larger crystals or snowflakes, each with a unique pattern. Much rain starts as snow, but melts before it reaches the ground.

Hail
A hailstone forms when strong air currents carry an ice crystal up and down again and again through a thunder cloud. The crystal collects many coatings of ice before it is heavy enough to fall.

Beaufort's Scale

In 1805, British admiral Sir Francis Beaufort (1774–1857) devised a scale for describing wind strength at sea, based on things such as wave height. This proved so successful that it was later adapted for land, using such indicators as smoke plumes and trees. In 1906, wind meters made it possible to relate the scale to specific wind velocities. Beaufort's Scale begins at Force 0, for calm, when smoke rises vertically and the sea is as smooth as glass. Force 12 denotes hurricane conditions.

Force 1
Calm. Trees are quite still. Wind speed 0-0.5 mph (0-1 kph).

Force 4
Moderate breeze. Small branches move. Wind speed 13-18 mph (20-29 kph).

Water in the air

There is enough water vapor in the troposphere to flood the world to a depth of a yard. But you can see it only when the air gets cold enough for the vapor to condense into water droplets, forming clouds or mist. Clouds form wherever air rises because the troposphere gets colder the higher you go. There are many ways air can rise, so there are many kinds of cloud. Fluffy white "cumulus" clouds, for instance, tend to appear on sunny days, because some spots are warmed more by the Sun than others, creating local updrafts called thermals. If air rises high enough, the water turns into ice, forming feathery "cirrus" clouds. This often happens along a warm front, where a mass of warm air meets a mass of cold air, and augurs bad weather.

Frost

When temperatures drop below freezing, water vapor in the air leaves a frosting of ice crystals over the ground. If it is the air that is cold, the frost is called "rime"; when it is the ground, "hoarfrost."

Fog and mist

Sometimes, when the air cools down quickly, or when a warm, moist wind blows over cool water, water vapor in the air condenses into a cloud of water droplets near the ground to form mist.

EXPERIMENT
Making a thermometer

Adult supervision required.

Most thermometers contain a fluid that expands or contracts according to the temperature — typically mercury or alcohol. This simple version uses colored water. Evaporation of the water makes readings unreliable, but it can give you at least a rough idea of the temperature.

YOU WILL NEED
● *glass bottle with stopper* ●
glass tube ● *beaker* ● *ink* ●
water ● *ice*

1 Fill the bottle with water colored with ink. Drill a hole in the stopper to fit the tube. Slide the tube just through the hole gently. DO NOT PUSH HARD. Smear the tube with petroleum jelly to help it slide through.

2 Place the bottle in a glass beaker of cold, iced water; reopen it and top it with cold water to the brim. Cut a piece of card the same height as the tube above the stopper, and tape it to the glass tube. Mark the level of water in the tube.

3 Make a scale by marking the level of water in the tube compared to temperatures on an indoor thermometer. Or just use it to show which way the temperature is changing.

Force 6
Strong breeze. Large branches sway. Wind speed 25-31 mph (40-50 kph).

Force 7
Near gale. Whole trees sway. Wind speed 32-38 mph (51-61 kph).

Force 9
Strong gale. Branches are blown down. Wind speed 46-54 mph (75-87 kph).

Wind station

WATCHING TREES SWAYING or flags flapping on a flag pole can give only a rough idea of how the wind is blowing. With this simple wind station, you can make fairly accurate measurements of wind strength and direction, and keep a record throughout the year. A simple weather vane indicates wind direction; a swinging arm and cup show wind strength.

Arrowhead indicates where the wind is blowing from

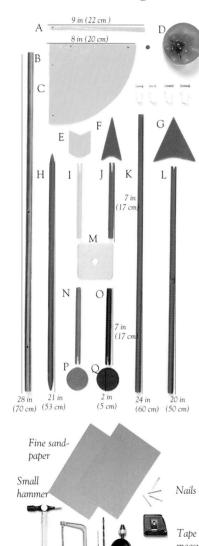

A 9 in (22 cm)
8 in (20 cm)
B
C
D
E F G
H I J K L
7 in (17 cm)
M
N O
7 in (17 cm)
P Q
28 in (70 cm) 21 in (53 cm) 2 in (5 cm) 24 in (60 cm) 20 in (50 cm)

Fine sand-paper

Small hammer

Nails

Tape measure

Hand files

Hacksaw Hand drill

Wood glue

Paint-brush

Turpentine Paints

EXPERIMENT
Making a wind station

Adult supervision is required.

YOU WILL NEED
- *plywood 1 x 1 ft (30 x 30 cm)*
- *7 ft 8 in (235 cm) of ¹/₂ in (12 mm) dowel*
- *copper tubing 28 in (70 cm) of ¹/₂ in (12 mm) diameter* ● *4 fasteners* ● *nuts and bolts* ● *nails* ● *wood glue* ● *tools* ● *paints* ● *plastic cup or base of a soft-drink bottle*

1 Cut the plywood (C, E, F, G, M, P, Q) and dowel (H, I, J, K, L, N, O) to size as shown. Cut slots in each end of the dowel — except for L, which should have a slot at one end and a groove in the middle.

2 Slot G into the end of L, and glue in place. Nail and glue K to the groove of L.

3 Drill two holes in each of the straight sides of C. Position C in the right angle created by K and L, and secure it with the four fasteners and the nuts and bolts.

4 Drill a horizontal hole through dowel L just behind the upright K, and another through A. Fix A to L with nut and bolt, so that A swings freely. Pin the cup (or bottle base, D) to A to complete the wind meter.

5 Sharpen ends of H to a point, and slot the end just inside tube B. Drill through both, and secure with nut and bolt.

6 Assemble the wind vane by slotting E into I, F into J, P into N, and Q into O, and glue to secure. Then attach the other ends of I, J, N, and O to M. Fix this direction pointer to the top of B.

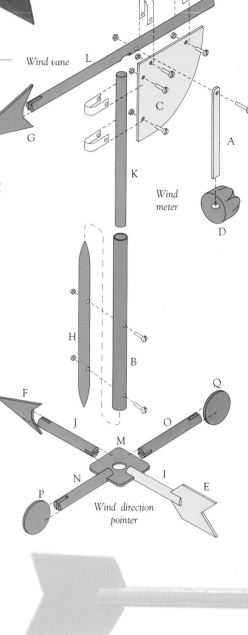

Wind vane L

Wind meter

G

C

A

K

Wind meter

D

H

B

F Q

J O

M

N I

P E

Wind direction pointer

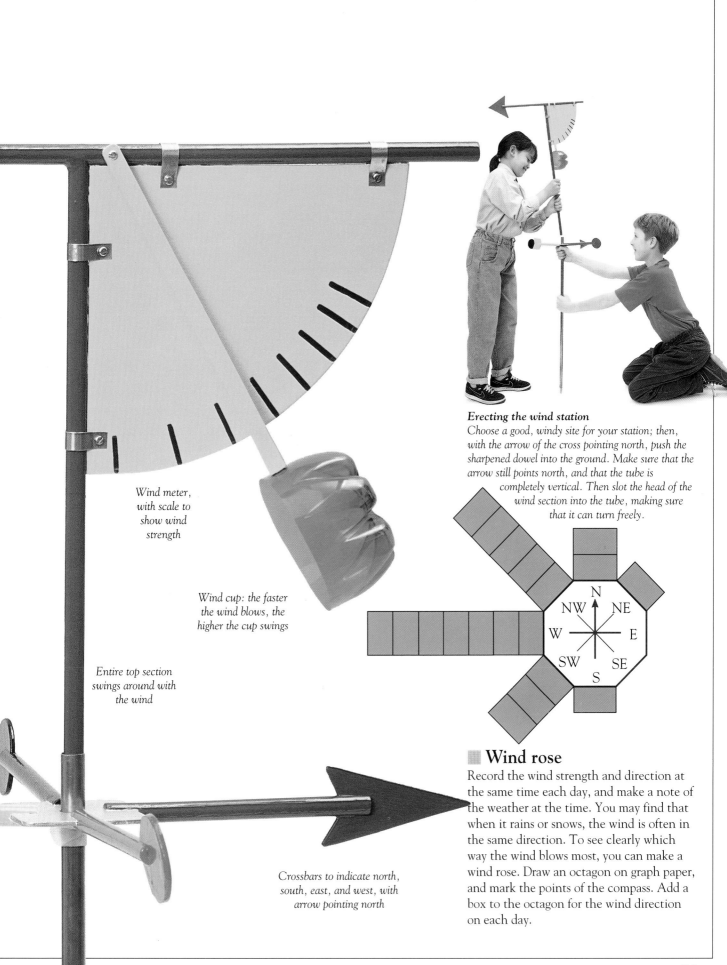

Wind meter, with scale to show wind strength

Wind cup: the faster the wind blows, the higher the cup swings

Entire top section swings around with the wind

Crossbars to indicate north, south, east, and west, with arrow pointing north

Erecting the wind station
Choose a good, windy site for your station; then, with the arrow of the cross pointing north, push the sharpened dowel into the ground. Make sure that the arrow still points north, and that the tube is completely vertical. Then slot the head of the wind section into the tube, making sure that it can turn freely.

Wind rose

Record the wind strength and direction at the same time each day, and make a note of the weather at the time. You may find that when it rains or snows, the wind is often in the same direction. To see clearly which way the wind blows most, you can make a wind rose. Draw an octagon on graph paper, and mark the points of the compass. Add a box to the octagon for the wind direction on each day.

Air streams

HOLD A SHEET of tissue paper between the fingers of both hands, and blow hard over the top. You will see something quite surprising. Instead of hanging limply, the far end of the paper begins to lift into the air. Indeed, the harder you blow, the higher it lifts. This is because of an effect known as the Bernoulli effect, after the Swiss mathematician Daniel Bernoulli, who first discovered it in 1738. Bernoulli observed that whenever air moves, its pressure drops. In fact, the faster air — or liquid — moves, the more pressure drops. So as you blow a stream of air rapidly over the paper, pressure here falls. Below the paper, air is still and pressure remains the same. The effect is to push the paper upward. The same effect enables birds and aircraft to fly and sailing ships to sail into the wind.

EXPERIMENT
The Bernoulli effect

Adult supervision is advised for this experiment

This simple experiment shows how pressure drops as air moves faster. It might seem strange at first that pressure should drop rather than rise. After all, moving things take more stopping. But each bit of air in a stream of air moves only because the air behind pushes it. So if the air is moving faster and faster, it can only be because the air behind is pushing harder — that is, its pressure is higher. So the pressure of fast-moving air must always be lower than that of slower or still air. The same is true for liquids flowing in channels or pipes.

YOU WILL NEED
● *two Ping Pong balls* ● *thread* ● *adhesive tape or glue* ● *drinking straw*

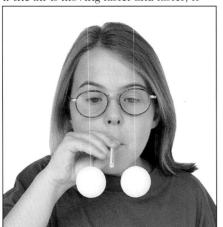

1 GLUE OR TAPE cotton threads to two Ping Pong balls. If you use tape, use as little as possible. Suspend the balls at equal heights roughly ³/₄ in (2 cm) apart. Now try to force the balls apart by blowing through a drinking straw as hard as you can.

2 YOU WILL FIND that the harder you blow, the closer the two balls swing together. This is because the harder you blow, the faster air streams between the balls. So pressure here drops, and the balls are pushed together by the greater pressure on the outside of each ball.

How a plane flies

It is because air pressure drops as it flows faster that wings can lift into the air anything from a paper glider to a jumbo jet weighing 350 tons. Wings provide lift because they cut through the air at a slight angle or are bowed upward. This means that air streaming over the top of the wing is forced to flow farther and faster, while air beneath flows more slowly. The air above the wing is thus at a lower pressure than the air beneath, and the pressure difference provides the lift.

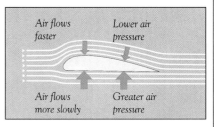

Air flows faster   Lower air pressure

Air flows more slowly   Greater air pressure

Wings provide lift because their curved shape forces air faster over the top than underneath.

Wings can provide lift only while they are slicing quickly through the air. If they move too slowly, the aircraft "stalls" and falls from the air. Gliders keep moving by flying downward, using rising air currents every now and then to give them extra height. Powered aircraft have jets or propellers to thrust them forward. Climbing demands far more thrust than level flight, and aircraft engines have to work hardest at takeoff.

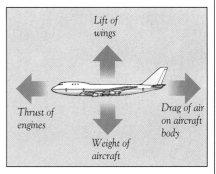

Lift of wings

Thrust of engines

Drag of air on aircraft body

Weight of aircraft

Air forces
To fly, an airplane's engines must thrust it forward fast enough for the wings to create the lift to overcome the plane's own weight and the drag of the air. So planes are made of light materials and streamlined to minimize drag.

EXPERIMENT
Streamlining and drag

 Adult supervision is advised for this experiment

To keep drag (friction with the air) to a minimum, aircraft, fast cars, and even birds are specially shaped, or "streamlined." The aim is to get air to flow around them as smoothly as possible, for the more the air is churned up, the greater the drag. One of the best shapes is a slim teardrop (like a fish).

This not only parts the air cleanly in front, but is tapered at the rear to let the air come together again with the least disturbance. You can see how little this shape upsets the airflow with this simple experiment with a candle flame.

YOU WILL NEED
● *candle* ● *saucer* ● *modeling clay* ● *cardboard* ● *paper clips*

1 HOLD THE CARDBOARD flat in front of a lighted candle, and persuade a friend to blow toward it. You will see that the flame gutters toward the cardboard. This is because the air stream curls and eddies around the card. The drag on a shape that disturbed the air this much would be considerable.

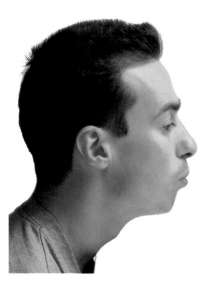

2 BEND THE CARDBOARD into a teardrop shape, clip it together, and line it up in front of the candle. When your friend blows this time, you will see the flame bend over smoothly, showing that the cardboard has upset the air stream very little. So it is a good, streamlined shape.

EXPERIMENT
Steady flow

Float a Ping Pong ball on the draft from a hair drier. The ball always sits right in the middle. This is another demonstration of the Bernoulli effect.

YOU WILL NEED
● *hair drier* ● *Ping Pong ball*

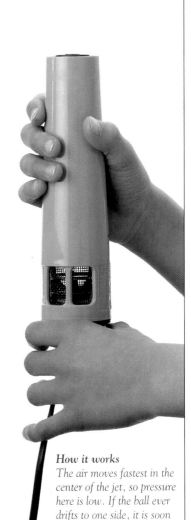

How it works
The air moves fastest in the center of the jet, so pressure here is low. If the ball ever drifts to one side, it is soon pushed into the middle again by the higher pressure where the air is slower at the edges of the air stream.

WATER

WATER IS A TRULY REMARKABLE SUBSTANCE. It is the most common compound on Earth, occurring as anything from tiny dewdrops to vast oceans. It is the basis of all life on Earth — life originally emerged from water, and water still plays a major role in all living processes. And its special nature means that it plays a part in a vast number of chemical reactions, from rusting to nuclear power production.

Less than 3 per cent of the world's water is freshwater, running in rivers, filling ponds and lakes, frozen in ice, and trickling through the ground. But even fresh–water is far from pure, containing a wide range of dissolved substances.

One of the unusual things about water is that it commonly occurs in all three phases of matter — that is, as a solid, a liquid, and a gas. Few other substances are ever naturally found in more than two states.

About 2 per cent of the world's water is solid — either frozen briefly as snow and frost or more or less permanently in glaciers and the polar ice sheet. Most of the rest is liquid — filling oceans, seas, and lakes, flowing in rivers and streams, or trickling through the ground (ground-water). Nearly three-quarters of the world's surface is covered in water.

The pressure of water increases dramatically with depth (p. 134). So deep-sea divers have to take special precautions to protect them from pressure effects.

■ Water vapor

Barely one thousandth of a per cent of the world's water is gaseous: that is, water vapor. But water vapor in the atmosphere plays a very important part in our weather. It is the presence of water vapor that distinguishes the "troposphere," the lowest layer of the atmosphere, from the atmosphere above. The troposphere is the layer in which the world's weather occurs, and in which we live, breathe, and move.

The amount of water vapor in the atmosphere is known as the "humidity." At any one temperature, air can only hold a certain amount of water vapor. Water constantly evaporates from seas, lakes, and the surfaces of plants, adding to the vapor in the air. When the air can take up no more water vapor, it is said to be "saturated." Once it is saturated, evaporation stops and droplets of water condense from the air, forming clouds or dew, and eventually falling to the ground as rain or snow (pp. 124-125).

■ Relative humidity

When scientists or meteorologists talk of "absolute humidity," they mean the total amount of vapor there is in the air. But the amount of vapor a particular volume of air can hold — that is, the saturation point — varies with temperature. The warmer air is, the more water vapor it can hold. So meteorologists talk also about "relative humidity." This is the amount of water there is in the air in relation to the most it could hold at that particular temperature.

Some substances are very good at absorbing water vapor from the air. Such substances are said to be "hygroscopic." Silica gel

Hygrometers (p. 141) show the acidity of water.

absorbs moisture so well that grains of silica gel are often put in with electronic equipment to keep the atmosphere perfectly dry. Hygroscopic substances like silica gel get damp when they absorb water vapor, but they do not dissolve. Any substance that dissolves in the water vapor it absorbs, like calcium chloride, is said to be "deliquescent."

■ What is water?

For thousands of years, scientists thought of water as an element, as the Greek philosopher Aristotle had averred in the 4th century B.C. But in 1784, the English chemist Henry Cavendish (1731–1810) conducted an experiment with hydrogen. He exploded mixtures of hydrogen and air with an electric spark. The reaction produced no change in weight but left him with what seemed to be pure water, suggesting that water was a compound including hydrogen.

A few years later, French chemist Antoine Lavoisier (1743–94) showed that water was a simple compound of hydrogen and oxygen. This can easily be proved by electrolysis (p. 156), which splits water into hydrogen and oxygen.

There is a special kind of water, called "heavy water," in which

Different liquids have different densities and float on each other (p. 140).

oxygen combines with deuterium instead of hydrogen. Deuterium is a special kind of hydrogen with an extra neutron in the nucleus of each atom (p. 27). Heavy water plays an important role in nuclear reactors, where it helps keep energy levels under control. It occurs in very small quantities in ordinary water, and can be extracted by distillation (p. 33) or electrolysis.

Water as a chemical

In every water molecule, there are two hydrogen atoms and one oxygen atom. The atoms are said to be bonded together "covalently" (p. 29). This means that the hydrogen and oxygen atoms "share" electrons. Hydrogen atoms have only one electron, so each hydrogen atom donates its electron to the oxygen atom and receives one electron from the oxygen in return.

The arrangement of these shared electrons has an important effect on water's chemical properties. The shared electrons are not located evenly within the molecule, but clustered toward one end, away from the center of the hydrogen atoms. Electrons are negatively charged (p. 148), so this makes one end of the molecule more negatively charged than the other. This difference in charge is called "polarity," so water is said to be a polar molecule.

One effect of this polarity is that the negative end of one water molecule is often

Ships float on seawater because they are less dense. They float slightly lower on fresh water, which is less dense than sea water (p. 139).

drawn toward the positively charged nucleus of a neighboring hydrogen atom — because unlike charges attract each other (p. 148). This is called a "hydrogen bond." Hydrogen bonds mean that water molecules often cluster together.

The attraction between water molecules creates enough surface tension to support a paper clip (p. 143).

Strange properties

It is this clustering of molecules that seems to explain some of water's rather strange properties. Going by other similar compounds, water should be a gas at room temperature, yet it quite clearly is not — although some water evaporates at almost any temperature above freezing. Water's boiling point 212°F (100°C) is far higher than that of any similar compound.

Water stays liquid because its molecules hang together so strongly that only a few molecules can escape to become gaseous. When it is frozen into ice, the molecules are organized into a regular, but widely spaced, lattice of crystals. But when ice melts, this lattice breaks down, and hydrogen bonding draws the water molecules much closer together. This is why, uniquely, water is denser when it is liquid than when it is solid, and why it expands, rather than contracts, as it freezes (p. 21).

The combination of water and air pressure can be used to make this fountain (p.136).

Water as a solvent

Another important effect of the water molecule's polarity is that it dissolves certain substances so well. Indeed, so many substances dissolve well in water that pure water occurs only rarely in nature. Sea water, for instance, is about 3.5 per cent salts, of which most is sodium chloride — the ordinary salt you use on your food.

Salts get into the sea from the rivers and groundwater that run into it, for even fresh water is very far from pure. Fresh water typically contains a range of dissolved substances including chlorides, carbonates and sulfates of metals such as sodium, magnesium, calcium, and iron.

When water is hard (p. 39), it is because of the magnesium and calcium compounds dissolved in it. Rainwater is fairly pure, but even rainwater tends to contain dissolved gases such as carbon dioxide and sulfur dioxide (which produces acid rain, p. 36).

It is because water is so good at dissolving things that it is vital for life. Life is thought to have begun in the oceans, which are actually very complex solutions. And all living organisms use "aqueous" (water-based) solutions, like those in the blood and digestive juices, to carry out biological processes. Of course, this is also why water pollution can be so dangerous. Pesticides, for example, can easily dissolve in water and be taken up by living organisms, from plants to human beings.

Waves are created by eddies in winds blowing over water. Just how big the waves grow depends on the strength of the wind and the "fetch" — that is, how far the wind blows over water. Over the vast expanse of the Pacific Ocean, surf waves can grow to vast size.

Liquids like water cannot be compressed (squeezed). So if confined within pipes and cylinders, liquids can be used to transmit enormous forces — such as the force to operate this hydraulic (waterlike) truck lift.

Water supply

MODERN CITIES NEED huge amounts of water. Every day, New York City, for example, consumes 1.8 billion gallons (7 billion liters). It would take 17 days to fill the city's 21 vast reservoirs, even using the entire cascade of Niagara Falls, a quarter of a million gallons a second (1 million liters). All this water is drawn from rivers, lakes, and wells, and pumped through a vast network of pipes to individual faucets. But whatever the source, water has to be cleaned before being fit for use. So it is usually pumped first into reservoirs, where solid debris settles to the bottom. Sluice gates let out water from the top into treatment plants. Here it is sieved to remove algae and trickled through gravel and sand to filter out smaller impurities. After further treatment with chlorine to kill off germs, the water is pumped into the main supply pipes (the "mains") or into short-term storage reservoirs.

Water treatment
After straining, water passes into filter beds — tanks containing layers of sand and gravel that are cleaned daily. Of course, these beds are much thicker than our homemade filter (see below) and produce much cleaner water!

EXPERIMENT
Cleaning water

This simple experiment shows just how effective even a crude homemade filter can be at removing dirt and debris from water. If you draw a bucket of water from a pond or the bed of a stream and then let it settle, you will see just how dirty untreated water is. Debris such as stones, soil, leaves, and sand is clearly visible, but there is also dirt you cannot see, such as bacteria and decaying life.

YOU WILL NEED
● *cupful of coarsely broken charcoal* ● *cupful of rinsed sand* ● *cupful of washed gravel* ● *6 in (15 cm) clay flowerpot* ● *coffee filter paper* ● *jug of pond water* ● *fine sieve* ● *large dish*

1 WASH THE CLAY POT WELL, and leave it to dry. Then line the pot with filter paper, place it in a dish, and pack one-third full with charcoal. Rinse the sand in a sieve under running water; while it is still wet, pack into the pot to fill the next third. Finally, wash gravel in a basin and use it to fill the pot.

Millions of water drop-
lets form a cloud.

The water cycle

However much water we consume, the
total amount of water in the world never
changes. This is because virtually all the
world's water is in the oceans, locked up
in the polar ice sheets, or involved in a
continuous cycle, called the hydrological
cycle, which is forever circulating water
between the sea and the sky. We get
most of our water by tapping into this
cycle. When your shower or bath water
runs away, it eventually ends up in rivers,
lakes, or the sea. From here it evaporates
in the heat of the sun to fill the lower
layers of the atmosphere with invisible
water vapor. A little of this vapor may be
carried aloft by rising air currents until it
cools enough to condense into clouds of
water droplets and ice. Once these grow
large enough, they fall as rain and snow
back to the Earth's surface, where some
runs down to the sea in rivers and lakes.
Some is trapped to fill reservoirs and
supply taps. So the cycle begins again.

*Water vapor
rising into
cold air condenses
into tiny drops of
water. If the air is
very cold, it may
turn into ice.*

*The Sun's heat
evaporates water
from both land
and sea, turning
it into invisible
water vapor.*

*Rivers
return
water to
the sea.*

*Droplets inside the
cloud join together
to form bigger
drops. When they
are large and heavy
enough, they fall as
rain or snow.*

*Some rain runs off
the land and into
streams and rivers.*

*Rivers may be
dammed to fill
reservoirs.*

*Some water seeps
away underground
through porous rock
to feed rivers from
springs.*

2 HOLD THE SIEVE over the pot, and gently pour in the pond
water in a steady stream so that you do not disturb the
layers of charcoal, sand, and gravel.

3 LARGE DEBRIS not caught by the sieve is trapped by the
gravel, while the sand traps the smaller pieces of dirt and
the charcoal and paper filters out the finest particles. If you
compare the pond water with the water seeping out into the
dish, you will see that it is much cleaner, although by no
means fit for drinking yet, for it contains invisible germs.

Water power

IF YOU HAVE EVER BEEN BOWLED OVER by a big wave in the sea, or even tried putting your finger under a gushing tap, you will know water can move with enormous power. Sometimes this power comes just from the water's momentum (p. 58) — once a heavy mass of water is moving, it is hard to stop. Waves are like this, set in motion by the wind and expending their energy only when they break on the shore. Sometimes, as with water running downhill in rivers and streams, the power is due to gravity. Sometimes the power comes simply from the water's depth, for, just as with the atmosphere, pressure in water — its power to push — increases with depth, as the experiment below demonstrates. People began to use water power thousands of years ago; the Romans used horizontal paddle wheels turned by streams to grind grain. Now, as the problems of using fossil fuels for power are becoming clear (p. 121), new ways are being sought to exploit the power of moving water.

EXPERIMENT
Water pressure

This simple experiment demonstrates how the pressure of water, or any liquid, increases with depth.

Water flows feebly from top hole

Water shoots out further from the middle holes because the pressure here is greater.

Jet from the bottom hole is longest of all, because pressure here is greatest.

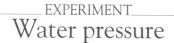

Adult supervision is required.

YOU WILL NEED
- *2-quart plastic drinking bottle* ● *dish*
- *nail* ● *water* ● *scissors*

1 Carefully off the top of the drinking bottlewith scissors, and make holes with a nail at four different levels.

2 Place the bottle in a dish, and cover the holes with your fingers. Ask a friend to fill the bottle with water. When it is full, take away your fingers. See how the lower holes shoot out longer jets of water because of the greater water pressure.

■ River power

Holding power
Dams may store water both for HEP and for consumption. Because water pressure increases with depth, dams must be thicker at the base or curved to spread the load.

Hydroelectric power (HEP) plants use moving water to turn turbines and generate 20 per cent of the world's electricity. They are often built beneath big dams because the dam creates a deep reservoir, and the pressure of this depth or "head" of water at the base of the dam turns the turbines very forcefully.

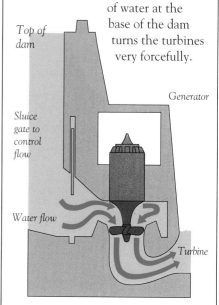

Top of dam

Sluice gate to control flow

Generator

Water flow

Turbine

When the gates open, water rushes through the base of the dam with great force, turning the turbine blades to generate electric power.

▉ Power from the waves

The need to find alternatives to fossil fuels has prompted research into many different sources of energy, including moving water. Hydro-electric power is clean, but building large dams can create enormous social and environmental problems. That is why scientists are now looking at the sea as an energy source. A few power plants already exploit the ups and downs of tides in the sea. But it is in the movement of ocean waves that the real potential of power from the sea seems to lie.

Various systems for harnessing wave power have been tried, and it seems likely that wave power could be not only safe and clean but also relatively cheap. However, the technology is at an early stage, and there remain considerable problems, not least of which is building systems to respond to small waves yet cope with storms.

Some wave-power generators are based on land, but scientists are seeking to develop methods that work out at sea, to exploit the ocean's full potential. One such idea is Stephen Salter's "Duck," so called because of its paddles, shaped like a duck, which bob up and down on the waves. As they bob, they turn a shaft to generate electricity. Another idea is rafts hinged to flap up and down with the waves, forcing water through pumps. People have even tried huge plastic bags containing air. Waves squash the air in the bags, and the moving air drives a turbine.

New wave
This is a prototype of a wave-powered generator built in a natural rock gully on the Isle of Islay, off the west coast of Scotland.

EXPERIMENT
Hydraulic power

Because a liquid cannot be squeezed, it can be used to transmit force through pipes. This experiment shows how hydraulic (liquid) systems, like car brakes, work.

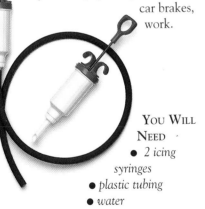

YOU WILL NEED
- *2 icing syringes*
- *plastic tubing*
- *water*

1 Fill one syringe with water, and then insert the nozzle into the plastic tube. Press the syringe to fill the tube with water.

2 Half-fill the second syringe with water, and fix it to the other end of the tube. Make sure there are no air bubbles inside the tube or syringes.

3 Press one plunger in, and you will see that the other plunger moves out – – the force has been transmitted.

4 Try the experiment with syringes of different sizes if you can. The system works just like gears (p. 72). A smaller plunger at the far end moves farther but less forcefully; a larger plunger moves less distance but more forcefully.

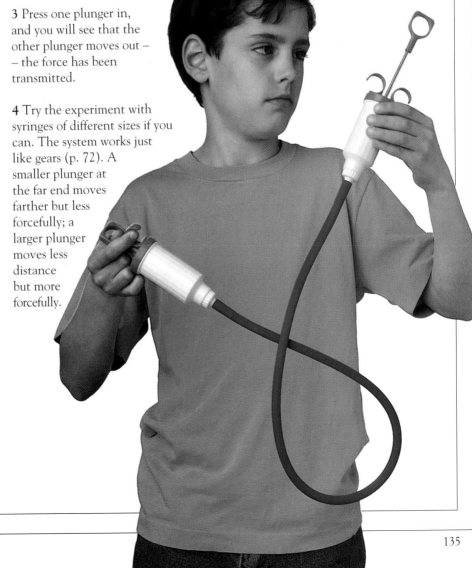

Raising water

As long as it is flowing downhill, water can be channeled where needed through pipes and aqueducts. This is why reservoirs are often sited on high ground. But when water has to be raised, or needs to flow faster — to keep a bathtub or sink from taking ten hours to fill, for example — siphons, pumps, and other methods must be used. Siphons raise water using nothing but the pressure of the water itself (p. 134). With pumps, you must provide some external force, whether a motor or just muscle. Pumps can both boost the power of the flow (water pressure) and lift water to great heights.

■ Hand pumps

Hand pumps have been in use since before 300 B.C. The simplest are the old-fashioned "lift" pumps once common in villages for drawing up water from underground. They lift a little water at a time with a plunger or piston moved up and down by a hand lever. Since they rely on air pressure to push water in behind the plunger, they can only lift water about 28 ft (9 m) or so. Most motorized pumps use rotating gears or vanes instead of a plunger to move the water, so they can pump water very quickly from great depths.

EXPERIMENT
Pressure fountain

This simple project uses the siphon principle to create a fountain that works by air pressure alone.

You Will Need
● *three glass jars, one with a cork stopper*
● *two straws*
● *modeling clay*
● *water colored with food coloring*

1 Drill two holes in the stopper, and push the straws into these holes. Fill one jar with about 2 in (5 cm) of colored water, and push the stopper into the jar. Adjust the height of the straws so that one is just sticking through the lid and the other is sticking halfway through. Seal around the holes in the stopper using the modeling clay.

2 Fill the second jar with colored water. Block the ends of straws with your fingers and invert the jar. Position it over the second jar so the straw that sticks halfway through dips into the water. Place the third jar under the other straw.

3 You will find that water rises up the straw from the second jar to form a fountain in the first jar. As water empties out of the first jar into the third jar, the air pressure in the first jar is reduced, causing water to be sucked up from the second jar.

EXPERIMENT
Making a siphon

Siphons have many uses in science and industry, and they are often used in the home for emptying tropical fish tanks. The siphon's gentle sucking action removes the water without disturbing gravel, snails, and water weeds.

You Will Need
● *two large glass bottles* ● *a piece of plastic tubing*

1 Place both bottles on a level surface. Fill one bottle three-quarters full of water. Place the plastic tube in the filled bottle, and suck water up the tube until it is filled.

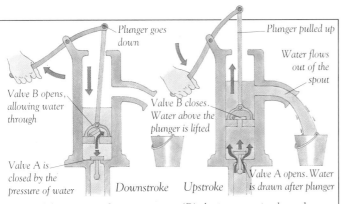

Plunger goes down

Plunger pulled up

Water flows out of the spout

Valve B opens, allowing water through

Valve B closes. Water above the plunger is lifted

Valve A is closed by the pressure of water

Downstroke

Upstroke

Valve A opens. Water is drawn after plunger

How a lift pump works

As the plunger rises, water is drawn in behind. As it goes back down, the pressure of water swings a valve (A) at the bottom shut and opens another

(B), letting water in above the plunger. When the plunger rises again, the pressure of water shuts valve A, so that water above the plunger pours out of the spout.

■ DISCOVERY ■
Archimedes' Screw

AROUND 2,200 YEARS ago, the Greek mathematician Archimedes invented a device for drawing up water from rivers to irrigate the land. Now called Archimedes' Screw, the machine consists of a cylinder containing a giant screw. Rotating the handle turns the screw and draws water upward.

A handle turns the screw

River

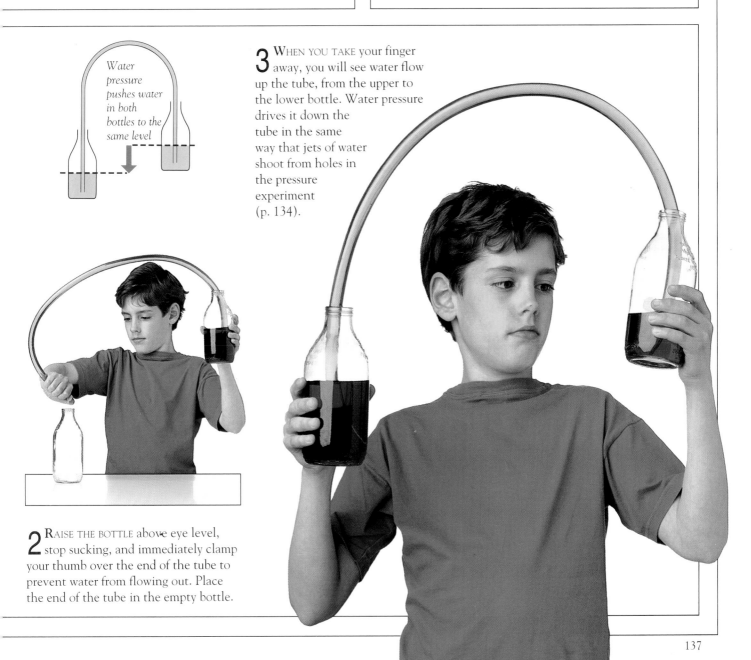

Water pressure pushes water in both bottles to the same level

3 WHEN YOU TAKE your finger away, you will see water flow up the tube, from the upper to the lower bottle. Water pressure drives it down the tube in the same way that jets of water shoot from holes in the pressure experiment (p. 134).

2 RAISE THE BOTTLE above eye level, stop sucking, and immediately clamp your thumb over the end of the tube to prevent water from flowing out. Place the end of the tube in the empty bottle.

Floating and sinking

WHY DO SOME THINGS FLOAT on water, and others sink? The answer is simple. It all has to do with density. Things that float are lighter or less dense than water; things that sink are heavier or more dense. When any object is submerged, its weight pushes down. But the water pushes back upward — with a force or "upthrust" equal to the weight of water displaced (pushed out of the way) by the object. If the object is less dense than water, the upthrust is enough to make it float to the surface; if it is more dense, it will sink. Wood floats because it is less dense than water. Ships float because of the air inside their hulls.

You can float on any water, but the dense, very salty water of Israel's Dead Sea makes people float especially high.

EXPERIMENT
Making things float

Any substance that is denser than water will sink. Stone, metal, and modeling clay are all denser than water and so sink. They sink because they weigh more than the upthrust provided by the water. This is because they are heavier than the water they displace when submerged. But dense substances can be made to float by reshaping them to increase the volume of water they displace, increasing the upthrust of the water. In this experiment, a ball of modeling clay that sinks is reshaped into a boat that floats!

1 P**OUR WATER INTO** the container until it is about three-quarters full. Make a compact ball from the clay, and drop it gently into the water.

2 T**HE BALL SINKS** because it is denser than water. Notice how the water level rises. This is because the ball displaces its own volume of water.

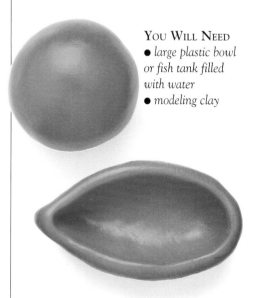

YOU WILL NEED
● *large plastic bowl or fish tank filled with water*
● *modeling clay*

3 T**AKE THE BALL** from the container, and reshape it into a boat with tall sides. By increasing the volume of the object, you increase the water's upthrust.

4 T**HIS TIME** the modeling clay floats. The higher water level indicates that a greater volume of water is displaced by the boat and the air that it contains.

EXPERIMENT
Measuring buoyancy

Some 2,200 years ago, Archimedes, the Greek mathematician who lived in Sicily, made a famous discovery. He found that when an object is immersed in liquid, it weighs less than it weighs in air. This is why in the swimming pool you can often lift someone far too heavy to lift when you are out of the pool. Archimedes explained that this "buoyancy" was due to the upthrust of the water, pushing the object up.

For something to float, its weight, which pulls it downward, must be balanced by the upthrust of the liquid. A floating object displaces a volume of liquid. Archimedes showed that the weight of the volume of liquid displaced by the floating object is equal to the upward force exerted by the liquid. Follow this experiment to measure the upthrust of water on a floating object and to prove that it is equal to the weight of the object.

YOU WILL NEED

- kitchen scales ● plastic jar ● jug ●
small floating object like a half-filled jar
- rectangular baking pan ● water

1 TAKE THE PAN off the kitchen scales, and adjust the needle so that it reads zero. Stand the scales in the cake pan, and put the plastic jar on the scales. Using the jug, fill the jar with water to the brim. Make a note of the weight.

2 GENTLY DROP THE OBJECT into the water. This will make some of the water spill out into the cake pan.

3 LOOK AT THE WEIGHT of the jar of water with the floating object. You will see that there is no change.

4 CAREFULLY LIFT THE JAR off the scales, and remove the cake pan. Replace the weighing pan, and reset to zero. Pour the water from the dish into the pan.

5 WRITE DOWN THE WEIGHT of the displaced water from the pan. Take the weighing pan off the scales, and adjust the needle to zero again.

6 WEIGH THE OBJECT. You will see that it weighs the same as the displaced water. The upthrust of the water is equal to the weight of the object.

Liquid density

JUST AS SOLID OBJECTS less dense than water float, so too do liquids that are less dense, providing they do not mix. A light liquid will float on top of a heavy one. If you look at puddles on a busy street, you will often see a thin layer of oil floating on the surface of the water. This is because oil dropped on the road by vehicles is less dense than rainwater. Indeed, most oils are lighter than water. In the same way, changing the density of a liquid, by either changing its temperature or dissolving things in it, will affect how well solid objects float in it. Boats float higher on salty seawater than fresh water because it is denser.

Slicks of spilled oil float on water, coating swimming birds.

EXPERIMENT
Floating liquids

You can show how liquids of different densities float on each other with this attractive demonstration of liquids floating in layers. It only works, of course, with liquids that don't mix, such as most oils and water.

YOU WILL NEED
● *tall glass* ● *syrup* ● *glycerol* ● *water colored with red food coloring* ● *olive oil or any light cooking oil* ● *rubbing alcohol colored with blue food coloring*

Starting with the heaviest liquid, syrup, pour it into the bottom of the glass. Then add the next heaviest, glycerol, dribbling it slowly down the inside of the glass so that the syrup is not disturbed. When this layer has settled, carefully add the water and then the olive oil. Finally trickle in the lightest liquid, the rubbing alcohol. You will end up with a striped effect in the glass.

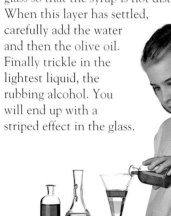

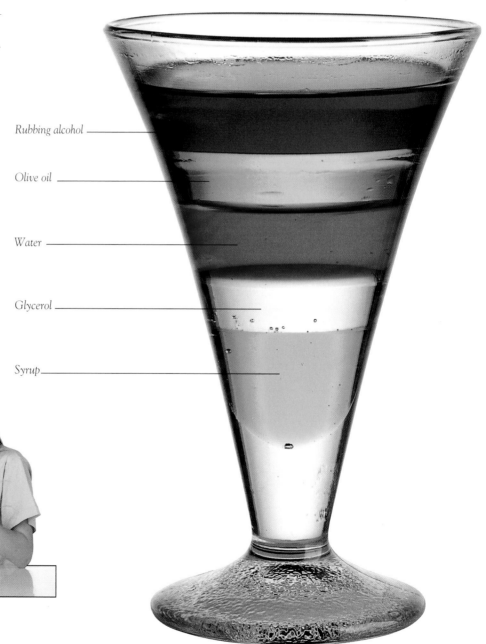

Rubbing alcohol

Olive oil

Water

Glycerol

Syrup

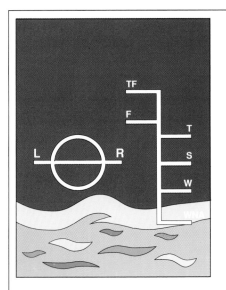

Plimsoll line
This mark on a British ship shows the level that the vessel should float in fresh water (F) on the left and salt water on the right: in the tropics (T), in summer (S), in winter (W), and in winter in the North Atlantic (WNA). Letters on ships of other nationalities may be different.

The Plimsoll line

BECAUSE WATER VARIES in density, ships can float at different heights in different water — even when carrying identical weights of cargo. Ships float higher in salt water than in fresh water, because it is denser. They also float higher in cold winter seas than in warm tropical seas, because the cold sea is denser. So when deciding just how much cargo to put on the boat, the captain needs to know how well the boat will float in the different waters the boat will sail through. If, for example, the boat is loaded to the limit in cold winter waters, it could well sink dangerously low when it sails into warm, less dense tropical seas. So most ships are marked with a scale on the hull called the Plimsoll line, made compulsory on British ships in 1876. This line shows the maximum safe level the ship can be loaded to in each type of sea.

Riding high
The passenger liner Queen Elizabeth II is floating so high in the water that the water level mark is clearly visible, showing that the ship is both unladen and in fairly cool, dense seawater.

Making a hydrometer

The simplest way to measure the density of a liquid is with a hollow rod, weighted at one end to make it float upright. It sinks deep in a light liquid but less deep in a heavier one. The level the hydrometer floats at is usually compared to pure water, giving a measure called relative density or specific gravity. You can make your own hydrometer like this.

YOU WILL NEED
● *3 glass beakers* ● *straw* ● *modeling clay*
● *water* ● *salt* ● *cooking oil* ● *marker pen*

Brewing beer
A hydrometer is used by brewers to measure the specific gravity of beer, indicating the sugar content (and so strength) of the beer.

Water

1 **P**OUR SOME WATER in one beaker. Stick a blob of modeling clay to the end of the straw, and float it in the beaker. Carefully mark on the straw the height that the water comes to.

Salt solution

2 **F**ILL THE SECOND BEAKER with salt solution, and float your hydrometer in it. You will see that it does not sink as low as it does in water.

Cooking oil

3 **U**SE COOKING OIL in the third beaker. This time the straw sinks lower than the water mark, as oil is less dense than water.

Surface tension

LOOK CLOSELY at the condensation on a really cold soda can or a glass of iced lemonade. All but the biggest droplets are perfectly round. So too are dewdrops on a spider's web and drops from a slowly dripping faucet — although you need high-speed photography to see this. Small drops of water always tend to be round because of a phenomenon called surface tension. Surface tension occurs because the molecules in water attract each other. In the middle of a drop, molecules pull toward each other equally in all directions. But at the surface, molecules are only pulled into the water, for there are no molecules to pull in the opposite direction, so the water pulls its surface tight around it like a stretched skin. Surface tension is only strong enough to pull small droplets into balls, but it is the reason why the bristles of a paintbrush cling together as you pull it out of water. It is also why the water surface bows up in a "meniscus" when you fill a glass carefully to the brim.

Walking on water
The skin created by surface tension is strong enough to bear the weight of tiny insects like pond skaters, which skim quickly across the water.

EXPERIMENT
Leakproof fabric

Surface tension stops water from pouring through the tiny holes between the threads in woven fabrics such as cotton or gauze — which is why tents keep out rain so long as you do not touch the fabric and break the tension. This experiment shows how well surface tension prevents water leaking.

YOU WILL NEED
● *milk bottle* ● *piece of fine gauze bandage*
● *rubber band*

Fill the bottle with water, and fix the gauze over the end with a rubber band. Turn the bottle over in a sink. The water will not flow out because surface tension acts like a skin to stop the water escaping through the holes in the fabric.

EXPERIMENT
Floating metal

You can make a metal paperclip float on water with the aid of surface tension!

YOU WILL NEED
● *glass of water* ● *paperclip*

Fill a glass tumbler with water, and rest a paperclip on the surface of the water. The paperclip will float, supported by the "skin" of the water's surface. This may take several tries to work. Look for the skin stretching under the weight of the paperclip.

■ Dry water?

There are times when we wish water would behave in a "wetter" way. It can be frustrating when water seems to simply roll off greasy dirt on the floor or on a car instead of washing it away, or when garden sprays collect in drops on leaves instead of coating them evenly. Surface tension is to blame for this annoyance. Water molecules are attracted to each other more strongly than many other things and will not wet them unless the surface tension is broken. Grease in particular is water repellent, and water rolls off greasy surfaces or clings together in drops. Soap contains "wetting agent" to break the surface tension so that grease can be properly wetted. Similarly, photographers use a wetting agent to prevent water from drying in blobs on films after developing.

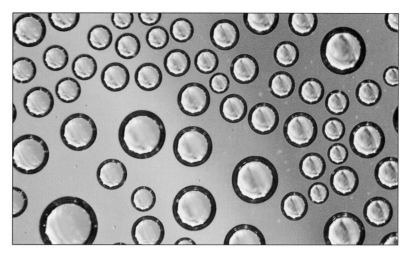

This is a magnified photograph of water on a sheet of clear plastic. Surface tension means the water does not spread evenly over the plastic but is pulled into almost spherical drops.

EXPERIMENT
Breaking the tension

This experiment shows just how effectively detergent breaks surface tension. Floating matches held in position by surface tension are instantly dispersed when a drop of detergent is added to the water, breaking the tension where the drop falls.

You Will Need
- *4 matches* ● *shallow dish of water*
- *dishwashing liquid* ● *dropper*

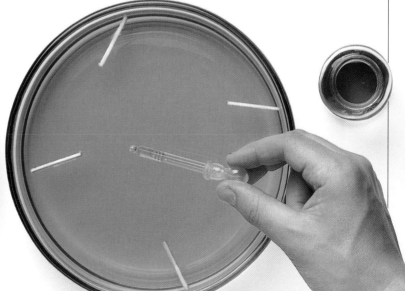

1 FILL THE DISH WITH CLEAN WATER, and allow it to settle so that the surface is completely smooth. Carefully float the matches on the surface of the water, arranged into a star shape as shown.

2 USING THE DROPPER, add a drop of dishwashing liquid to the water in the center of the dish. This breaks the tension there, and the matches are instantly drawn outward by the stronger surface tension around the edge of the dish.

Electricity

■■■■■■ AND ■■■■■■

magnetism

ELECTRICITY IS NOT JUST the energy
that makes everything from toasters to
televisions work. Nor is magnetism
just the invisible impulse that makes
compasses spin. Together, electricity
and magnetism are one of the
fundamental forces of the universe.

*There are few more dramatic demonstrations of electrical energy than
bolts of lightning — static electricity that jumps between thunderclouds
and the ground with such enormous power that it heats the surrounding
air ferociously. Thunder is the crash of the rapidly expanding hot air.*

ELECTRICITY

Eʟᴇᴄᴛʀɪᴄɪᴛʏ ɪꜱ ᴏɴᴇ ᴏꜰ ᴛʜᴇ ꜰᴜɴᴅᴀᴍᴇɴᴛᴀʟ ꜰᴏʀᴄᴇꜱ that hold all matter together. It is also the most versatile of all forms of energy. It can provide the heat to make lightbulbs glow and electric heaters warm. It can generate the sound in a radio and the picture in a television set. And it can power anything from the thousands of minute microswitches in a computer to the huge engines in a high-speed train.

18th-century American scientist and statesman Benjamin Franklin was the first to suggest the idea of positive and negative electrical charges. He also showed that lightning was static electricity by flying a kite in a thunderstorm — an experience he was lucky to survive.

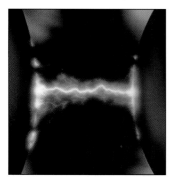

In the 1930s, American physicist Robert van de Graaff (1901–67) invented a machine able to produce giant sparks of static electricity.

Until the 18th century, few scientists took much interest in electricity. The Ancient Greeks were aware of some electrical effects. Around 600 B.C., for example, the Greek philosopher Thales observed that when he rubbed a kind of resin called amber with a cloth, it attracted feathers, threads, and bits of fluff. Indeed, the word "electricity" comes from the Greek word for amber, *elektron*. But few people considered it much more than a minor curiosity.

One early scholar who did investigate electric effects was Queen Elizabeth I of England's physician, William Gilbert (1544–1603). Gilbert showed that not only amber but glass and sulfur too gained this strange ability to attract light things when rubbed. It was Gilbert who decided to call this unknown force of attraction "electricity."

▓ Electric fluid

For a century and a half, few other scientists bothered much about electricity. Then, in 1733, the French chemist Charles Dufay (1698–1739) found that while some objects attract each other when rubbed, similar objects seemed to repel. If two amber rods were electrified by rubbing, for example, they seemed to repel each other. Dufay decided electricity is a kind of fluid and that there must be two kinds of electricity. "Vitreous" electricity, he decided, is released when you rub glass, crystal, hair, or wool; "resinous" electricity is generated in amber or silk. Unlike kinds attract each other; like kinds seemed to repel.

▓ Conductors

At around the same time, the English physicist Stephen Gray (1666–1736) showed that some substances conduct electricity and some do not. Indeed, he found he could transmit electricity created by rubbing a glass tube more than 330 ft (100 m) along a silk thread. Metals and other substances through which electricity passes easily came to be called "conductors"; glass, amber, silk, wood, and other substances that tend to block the path of electricity were called "insulators." In a famous demonstration, Gray showed how well the human body conducts electricity by suspending a poor charity boy from silk threads and connecting him to a static electric charge. The charge flowed through the boy's body so that he could lift scraps of paper from the ground without touching them.

▓ Opposite charges

It was the ingenious American statesman and inventor Benjamin Franklin (1706–90), however, who showed that electrical effects are much more than simply party tricks. Like Dufay, Franklin believed electricity is a kind of fluid, but he was convinced that there was just one fluid, not two, involved. When glass is rubbed, he thought, electric fluid flows into it, making it "positively charged"; but when amber is rubbed, electric fluid flows out of it, making it "negatively charged." Whenever they came into contact, Franklin believed, fluid would flow from positive to negative until it was equally balanced.

In fact, this is remarkably near what we now know to happen — except that it is tiny subatomic particles called

Electroscopes detect a static electric charge and show whether it is positive or negative.

electrons (p. 27) that are involved, and static electricity discharges from the amber to the glass, not vice versa. When amber is rubbed with a cloth, for instance, or when you pull a comb through your hair, electrons from the cloth rub off onto the amber, or from the comb into your hair. Electrons are said to be negatively charged, so the rubbed amber too is negatively charged. Since the cloth loses electrons, it becomes positively charged.

By the mid-1700s, there were machines able to generate quite large electrical charges when a handle was turned to rub glass or sulfur. Scientists could even store an electrical charge by insulating it in a special kind of glass jar, called a Leyden jar. Electricity could be drawn from the jar via a brass chain in the stopper. The charge could be so big that scientists could make brilliant sparks jump from the chain to their hands. Several got unpleasant shocks.

■ Lightning

Seeing these sparks, Franklin began to wonder whether lightning might be exactly the same thing. To test his idea, Franklin flew a kite in a thunder storm, and attached a key to the string on a short silk thread. Electricity flowed down from the clouds through the wet string and the silk to the key. When Franklin put his hand near the key, he felt a shock and saw sparks, exactly like those from the Leyden jar. He had proved his point, but was lucky to be alive. Many repeated his experiment and were electrocuted.

Within a few months, in 1752, Franklin devised a lightning rod – – an iron rod attached to the highest point of a building and wired to the ground. If lightning struck the building, it discharged harmlessly through the rod to the ground.

Franklin's discovery sparked off huge interest in electrical phenomena, and demonstrations of electrical effects became very fashionable. In 1786, the Italian anatomist Luigi Galvani (1737– 98) hung the legs of a

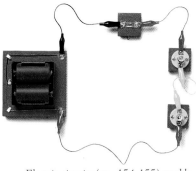

Electric circuits (pp. 154-155) enable electric energy to be used to operate anything from lightbulbs to computers.

dead frog on a railing in a thunderstorm to see if lightning would make them twitch as the spark from a Leyden jar had. To his surprise, they twitched even before the storm arrived, and Galvani believed he had discovered a new kind of electricity, made by the frog's nerves. He called this "animal electricity," and many people began to believe that this was the secret of life itself, the strange force that animated flesh and bone. Soon dozens of scientists were trying to bring corpses back to life by electrifying them. This is why, in Mary Shelley's famous story of *Frankenstein*, the mad doctor's monstrous creation is brought to life by a massive electric shock from a bolt of lightning.

This simple cell, one of the earliest batteries, works by chemical reaction (p. 151).

■ The first battery

Alessandro Volta (1745–1827), however, suspected it was not any "animal electricity" that was making the frog's legs twitch at all, and in the 1790s, he found the truth. The electricity came from the chemical reaction between the metal of the railing and the metal hook on which the legs hung.

Knowing this, Volta was able to make the first battery in 1800,

A reaction between carbon and the zinc casing creates a current in dry-cell batteries.

by building up alternate layers of copper and zinc in a jar of salt water. The chemical reaction created, for the first time, a steady supply of electricity.

Scientists soon learned a great deal about electricity. A crucial discovery was the electric circuit — if a conducting material is arranged in a loop, an electric charge flows through it from one terminal (one of the two metals) of a battery to the other. People likened this to the flow of water in a stream, and called it a "current." In fact, the analogy with water is misleading, for unlike water in a stream, the electrons that carry the electric charge barely move; they pass the charge on in relays. But the discovery of circuits meant that electricity could be fed through wires to where it was needed and, eventually, perform many different tasks.

In the 1820s, the French physicist André Ampère (1775– 1836) made many discoveries about the nature of electricity, while German physicist Georg Ohm (1787–1854) showed that the flow of an electric current through a wire depended on its "resistance."

However, it was not until the relationship between electricity and magnetism was fully appreciated, and it became possible to generate electricity at will (pp. 158-159), that electricity began to transform everyday life. Even then, scientists did not begin to understand just what electricity is, and its role in shaping matter, until the discovery of the electron in the 1880s (p. 15).

In 1800, months after Volta made the first battery, the British chemist William Nicholson (1753– 1815) showed that if leads from the battery were immersed in water, bubbles of oxygen and hydrogen were produced (pp. 156-157). This reaction, called electrolysis, is the basis of many important chemical processes.

Electrics have long played an important role in gas-engine cars, providing the spark to ignite the fuel, and powering items such as lights. Now electronic-control systems are playing an increasing role in keeping the car running efficiently. One day, many cars may even be powered by electric motors.

Static electricity

STATIC ELECTRICITY is the crackle when you comb your hair and the tingling when you take off a nylon sweater. The lightning that flashes in a thunderstorm is static electricity too, and so is the force that drags ink onto paper in a photocopier. What they all have in common is that they are produced by two things rubbing together. With the comb, it is the comb and your hair that rub; with lightning, it is ice crystals in a cloud. The effect is the same; rubbing produces a force called a "charge," which either attracts or repels. What happens is that when the two substances rub together, electrons are knocked off the atoms in one and stick to the atoms in the other. Substances that lose electrons are said to be positively charged; those that gain are said to be negatively charged. Unlike charges attract each other; like charges repel.

Sticky balloons
Rub balloons on a T-shirt or woolen sweater to give them a charge. A negative charge on the balloons makes them stick to the positively charged T-shirt.

Hair lift
Run a comb through your hair, and you can use it to pick up paper! This is because the comb gets a coating of negatively charged electrons that attract the paper.

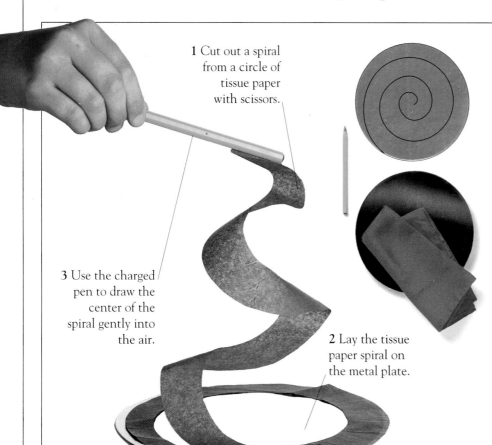

1 Cut out a spiral from a circle of tissue paper with scissors.

3 Use the charged pen to draw the center of the spiral gently into the air.

2 Lay the tissue paper spiral on the metal plate.

EXPERIMENT
Snake charming

Using static electricity, show how a charged object attracts an uncharged one. This works best in dry weather.

YOU WILL NEED
- *tissue paper* ● *silk handkerchief*
- *plastic pen* ● *metal plate or tray*

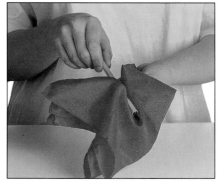

Charging the pen
Rub the plastic pen vigorously on the silk handkerchief, so that it becomes charged with static electricity.

■ Lightning strike

A bolt of lightning contains enormous energy — enough to power a small town for a year. Lightning starts inside thunderclouds, where air currents throw ice crystals and water drops together so hard that they become charged. Positively charged particles collect at the top of the cloud and negatively charged particles at the base. Eventually a lightning bolt — a stream of electrons – – leaps between the two to cancel the charge. This is sheet lightning. Dramatic forks are seen when lightning flashes not inside the cloud but toward the ground. Lightning discharges to the ground because the ground too has a slight charge. Forks take the shortest possible route to earth — which is why trees and tall buildings are so vulnerable to strikes.

EXPERIMENT
Making an electroscope

To measure the strength of a static charge, you can make a device called an electroscope. This experiment works best in dry weather.

YOU WILL NEED
● *small glass jar* ● *stiff wire* ● *aluminum foil* ●
thin candy-wrapper foil ● *cardboard disc* ● *plastic*
pen ● *cardboard* ● *tape*

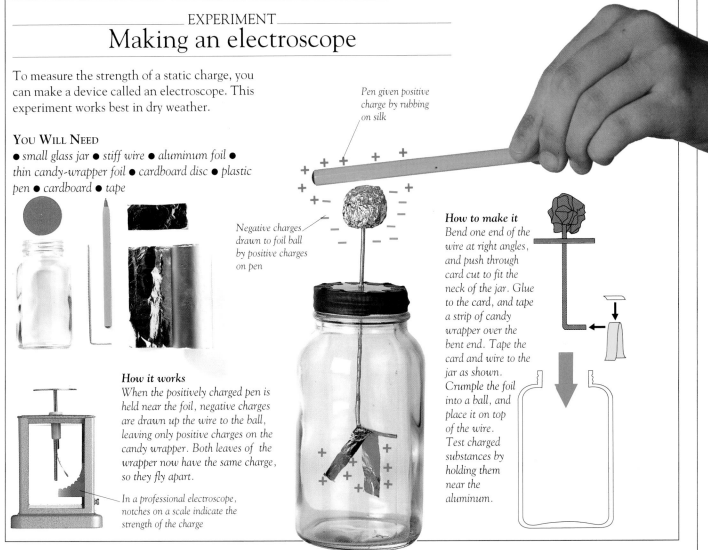

Pen given positive charge by rubbing on silk

Negative charges drawn to foil ball by positive charges on pen

How to make it
Bend one end of the wire at right angles, and push through card cut to fit the neck of the jar. Glue to the card, and tape a strip of candy wrapper over the bent end. Tape the card and wire to the jar as shown. Crumple the foil into a ball, and place it on top of the wire. Test charged substances by holding them near the aluminum.

How it works
When the positively charged pen is held near the foil, negative charges are drawn up the wire to the ball, leaving only positive charges on the candy wrapper. Both leaves of the wrapper now have the same charge, so they fly apart.

In a professional electroscope, notches on a scale indicate the strength of the charge

Current electricity

STATIC ELECTRICITY gets its name because the charge is static — that is, it does not move, or at best makes a single jump. Static effects like lightning are dramatic but hard to make use of. Yet an electrical charge can be made to move in a continuous stream, or "current," and then it becomes much more useful. Current electricity makes possible everything from electric light to supercomputers.

To make an electric current flow, two things are needed. First of all, there must be a continuous, unbroken path, or "circuit," for the current to flow through. This is usually a loop of copper wire because, like many metals, copper is a good conductor — that is, it lets an electric current pass very easily. Second, there must be a driving force (called the "electromotive force") to push the electrons that carry the charge around the circuit. The simplest way of providing this force is with a battery.

Christmas lights
Like all electric lights, these brilliant Christmas illuminations work because of electrons moving in an electric current around a circuit.

EXPERIMENT
Making a battery

Adult supervision is suggested. Batteries create a current chemically. Two "electrodes" (one copper, one zinc) are linked by wire and dipped into acid (the electrolyte). A chemical reaction in the acid leaves an excess of electrons on the zinc, creating a current that flows through the wire to the copper. In dry cells (e.g., flashlight batteries), the electrodes are the zinc case and a carbon rod. The electrolyte is a paste.

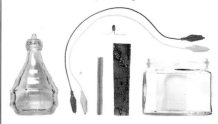

YOU WILL NEED
● wide glass jar ● strip of zinc ● copper pipe ● LED ● two leads ● vinegar

■ DISCOVERY ■
The first battery

CURRENT ELECTRICITY was discovered almost by accident. In the late 1700s, scientists could create static charges by rubbing, and Luigi Galvani could make the legs of dead frogs twitch with a charge. One day, however, Galvani made the legs twitch without any charge. It was Alessandro Volta who realized why: the brass hooks on which the legs hung were reacting chemically with an iron stand to create an electric current.

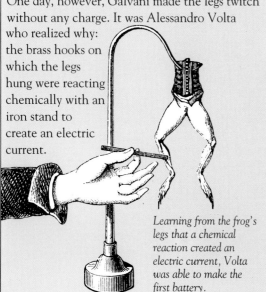

Learning from the frog's legs that a chemical reaction created an electric current, Volta was able to make the first battery.

■ What is current?

People once thought electricity flowed like water, which is why it was called an electric "current." In fact, it is more like a row of marbles — flick one end hard, and the marble at the far end shoots off. In any material that conducts electricity well, each atom has at least one electron held only loosely. Such "free" electrons normally drift at random from atom to atom. But if there are more electrons at one end of the circuit than the other, they push all the free electrons the same way, because like charges repel. This is what happens when you switch on an electrical circuit. No electron moves far; it simply knocks into the next one along the line, passing the current through the circuit.

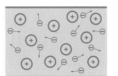

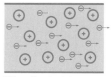

Free electrons (minus signs) jump between atoms at random

When the circuit is switched on, they all move the same way

Current always flows from negative to positive. With a battery, it flows one way all the time (direct current). With certain types of generator (p. 166), the terminals are continually switched over to create an alternating current, as shown here.

Electromotive force created by generator

Lightbulb

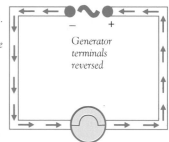

Generator terminals reversed

1 FILL THE GLASS JAR with vinegar. Vinegar is dilute acetic acid and forms the electrolyte for the battery.

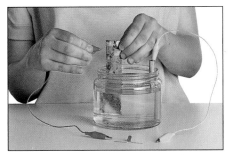

2 CLIP ONE END OF THE WIRES to the copper and zinc and the other to each terminal of a small lightbulb such as an LED.

Positive and negative

The chemical reaction strips positively charged zinc ions (p. 156) from the zinc strip, leaving it with an excess of negatively charged electrons. This excess sends an electric current through the wire to the copper pipe, which, in turn, draws positive ions from the acid and becomes positively charged.

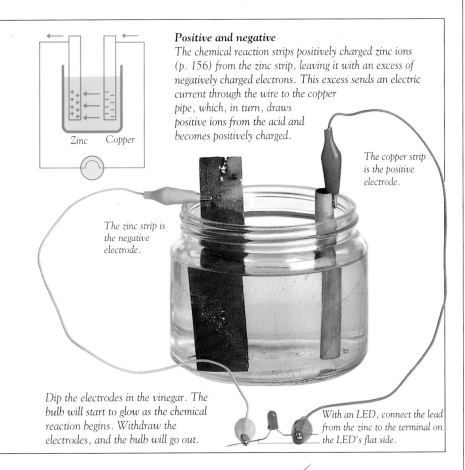

Zinc   Copper

The copper strip is the positive electrode.

The zinc strip is the negative electrode.

Dip the electrodes in the vinegar. The bulb will start to glow as the chemical reaction begins. Withdraw the electrodes, and the bulb will go out.

With an LED, connect the lead from the zinc to the terminal on the LED's flat side.

EXPERIMENT
Conductors

Adult supervision is suggested. Electricity travels through some materials easily (conductors) and some only with difficulty (insulators). Unlike insulators, conductors have many free electrons. Test materials to see whether they are conductors or insulators.

How to test for conductivity

Connect up the battery and bulb as shown. Then take one lead from the bulb to the test item and another from the far end of the test item to the free battery terminal. If the bulb lights up, the item is a conductor; if it does not, it is an insulator.

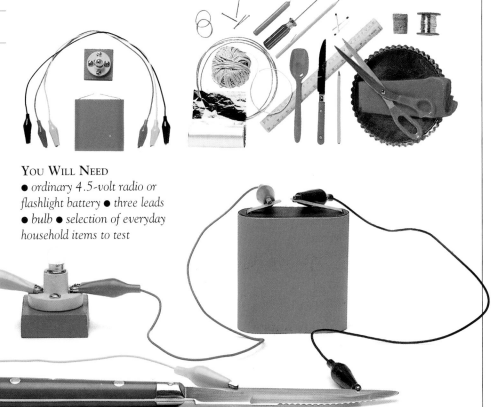

YOU WILL NEED
● *ordinary 4.5-volt radio or flashlight battery* ● *three leads* ● *bulb* ● *selection of everyday household items to test*

Making connections

THE WAY TO STOP an electric current is to break the circuit. Break the circuit, and the flow of electricity stops instantly. Similarly, the way to start an electric current is to complete a circuit. This is why nearly every electrical circuit has a switch, a simple device that either breaks the circuit or completes it to stop or start the current flowing.

Because electricity flows only if the circuit is a continuous, unbroken loop, all electrical equipment, whether just a simple lightbulb or a sophisticated hi-fi, must be connected into and made a part of this loop. When you switch on anything electrical, you are simply making the final connection in the electrical circuit. If

Spark plug
In a spark plug there is a gap in the circuit. But the voltage (pressure) of the electrical current is large enough for it to jump across the gap.

the equipment does not work, the chances are the circuit is broken.

Many circuits are made up of lengths of wire connected together. But there are other types of circuit. In a flashlight, for example, part of the circuit is often just a brass strip, while in equipment such as radios, it may be only a thin line of copper printed on a board. In many electronic items, circuits are just microscopic tracks formed on chips of silicon.

EXPERIMENT
Steady-hand game

Adult supervision is required. Current flows only when a circuit makes an unbroken loop. Test your hand control in this game. The hand held loop acts as a switch — the bulbs on the carrot light up when the loop touches the rabbit, making the final connection in the circuit.

YOU WILL NEED
● *chipboard* ● *balsa wood* ● *dowel*
● *battery* ● *two bulbs* ● *coat hanger wire*
● *metal strip* ● *connectors* ● *insulated wire*
● *tape* ● *cuphook* ● *saw* ● *drill and drill bits*

EXPERIMENT
Connecting up

Adult supervision is required. You can make a simple circuit using basic household items. Make a switch from a paperclip, and use it to control the lightbulb. When the paperclip touches both eyelets of the switch, the circuit is complete and the bulb will be lit. To turn the light off, break the circuit by pushing one end of the clip away from the eyelet.

YOU WILL NEED
● *batteries* ● *leads* ● *balsa wood* ● *bulb with holder* ● *screws* ● *screw eyes* ● *paperclip*

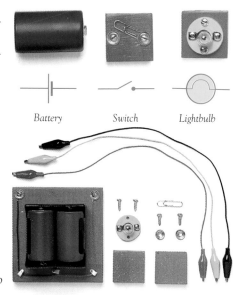

Battery   *Switch*   *Lightbulb*

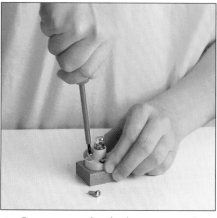

1 CUT A BLOCK for the batteries, and run wires from the positive and negative terminals. Fit the bulb into the holder, and screw it onto a balsa block.

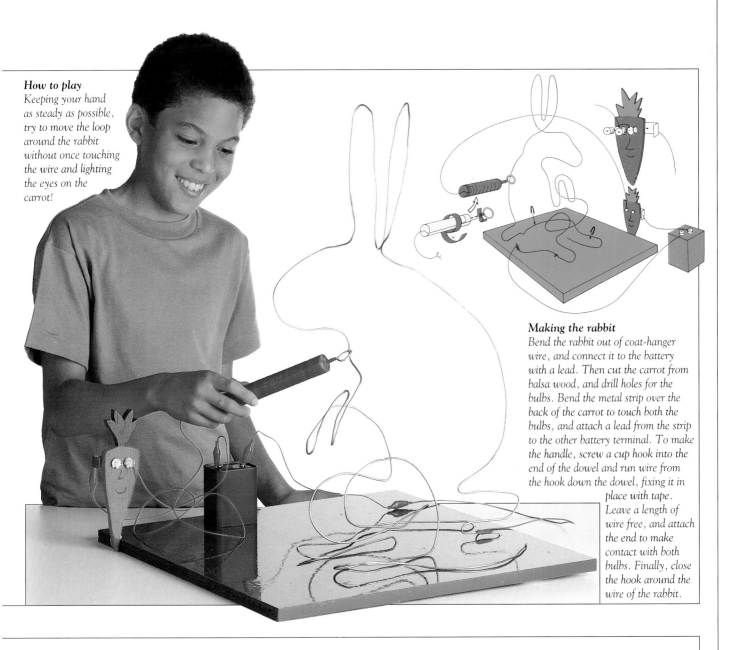

How to play
Keeping your hand as steady as possible, try to move the loop around the rabbit without once touching the wire and lighting the eyes on the carrot!

Making the rabbit
Bend the rabbit out of coat-hanger wire, and connect it to the battery with a lead. Then cut the carrot from balsa wood, and drill holes for the bulbs. Bend the metal strip over the back of the carrot to touch both the bulbs, and attach a lead from the strip to the other battery terminal. To make the handle, screw a cup hook into the end of the dowel and run wire from the hook down the dowel, fixing it in place with tape. Leave a length of wire free, and attach the end to make contact with both bulbs. Finally, close the hook around the wire of the rabbit.

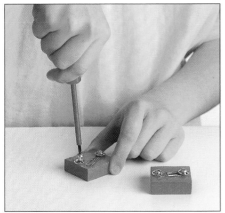

2 MAKE THE SWITCH by screwing a screw eye into a balsa block over one end of a paperclip. Fix a another eyelet to touch the other end of the clip.

When making a circuit, all connections must be good. Buy leads with alligater clips or strip 1 in of insulating plastic from the ends of the lead with scissors, taking care not to cut the wire. Twist the strands together with pliers, and screw down firmly.

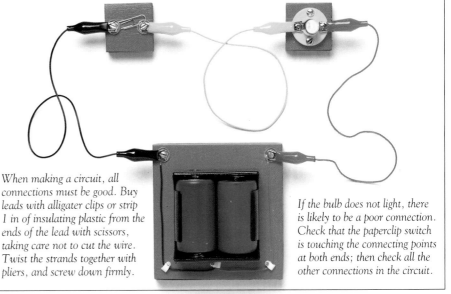

If the bulb does not light, there is likely to be a poor connection. Check that the paperclip switch is touching the connecting points at both ends; then check all the other connections in the circuit.

Simple circuits

CIRCUITS are connected up in many ways, but they all have three elements. There is a conductor through which the current flows; a load, which is the equipment being powered; and an energy source, such as a battery. Circuits need energy to drive the current, and the energy is basically the difference in the number of electrons at the negative and positive terminals of the battery. This difference is called the "potential difference" and is measured in "volts." The more energy a battery can give, the bigger the voltage. Flashlights typically use 1.5-volt batteries, while electricity in the home (p. 166) is at least 120 volts, depending on the country. But the rate at which the current flows — the "amperage" — depends not only on voltage but also on how well the circuit conducts electricity. This is called the "resistance" and is measured in "ohms" — after the scientist Georg Ohm, who found how voltage and current were related.

Edison's lightbulb

THOMAS EDISON'S LIGHTBULB, invented in 1879, was a simple idea that changed everyday life. Like modern bulbs, it relied on electrical resistance. Edison sealed a carbonized cotton thread in a glass bulb and pumped out the air. The thread, or "filament," was so fine that the current had to work hard to get through — so hard that the thread heated up and glowed brightly, giving out light.

Edison was one of the most successful inventors the world has ever known. Besides the lightbulb, he invented the phonograph and was a pioneer of the telephone. He became a rich man, but is said to have been more interested in his work than in money and often worked through the night on inventions.

Early bulb
Edison pumped all the air out of the glass bulb to ensure that the cotton filament did not burn up. Today the filament is usually made of tungsten, which glows much hotter and brighter, and the bulb is filled with an inert gas (p. 28).

EXPERIMENT
Varying resistance

Adult supervision is suggested. When electrons bump into atoms in the conductor, the current is reduced. This is the wire's resistance. The longer and thinner the conductor, the greater the resistance.

YOU WILL NEED
● *battery* ● *3 leads* ● *bulb* ● *soft pencil*

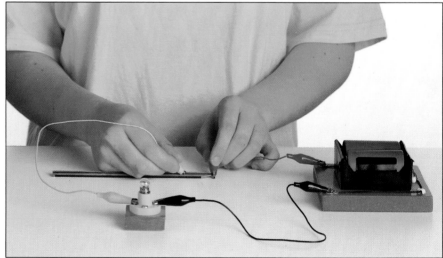

STRIP THE WOOD from the pencil after soaking in water. Attach the wires to the bulb and battery terminals, and connect the bulb and battery. Touch the pencil lead with the free ends, close together but not in contact. The bulb should light up. Move the wires apart so that there is more pencil lead in the circuit, increasing the resistance and making the bulb dimmer.

EXPERIMENT
Different circuits

Adult supervision suggested.
Electrical circuits must form a complete loop, but they can be connected up in two ways. If all the components are in a single loop, the circuit is said to be connected in "series." If the circuit splits into branches, it is connected in "parallel." A short circuit occurs when there is an easier alternative route — that is, with less resistance for the electricity to follow.

YOU WILL NEED
- battery ● leads ● switches ● bulbs
- three-way connectors

Series circuit
To make a series circuit, connect all the components — battery, switch, and bulb — in a single loop. The current should be exactly the same everywhere in the circuit.

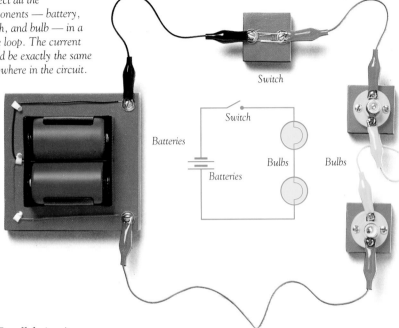

Switch

Switch

Batteries

Bulbs

Bulbs

Batteries

Parallel circuit
To make a parallel circuit, use connectors with three terminals to split the circuit into two (or more) branches, with one bulb on each. In the parallel circuit shown here, each branch takes exactly half the current. But if one branch is more resistant than the other — if you put a second bulb in one branch, for instance — less of the current will flow through this branch.

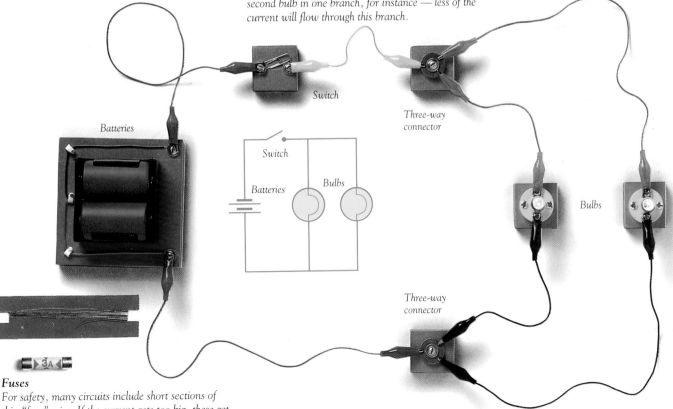

Batteries

Switch

Switch

Batteries

Bulbs

Three-way connector

Bulbs

Three-way connector

Fuses
For safety, many circuits include short sections of thin "fuse" wire. If the current gets too big, these get so hot that they burn through, breaking the circuit.

Electrical chemistry

IN EVERY BATTERY, the current is created by a chemical reaction. But the reverse can also happen; an electric current can spark off a chemical reaction. In fact, a current triggers a reaction in any electrolyte — that is, any liquid that conducts electricity. Most electrolytes are solutions of acids, bases, and salts that form "ions" (charged atoms) when dissolved in water. Yet even tap water reacts if you dip in two electrodes and switch on. When a current passes through an electrolyte, the electrolyte splits up chemically, at either the anode (the positive electrode) or the cathode (the negative electrode) or both. This process — called "electrolysis" — is used industrially both for extracting pure metals such as aluminum from rough ore and for electroplating (coating metals with such finishes as chrome and silver).

Zincing fast
During manufacture, most car bodies are now put through various treatments to protect them from rust. Among these is a dip in a zinc bath. The car body is electrified, and by a process of electrolysis, zinc is drawn from the bath and onto the body, quickly giving it a complete and even coating.

EXPERIMENT
Copper plating

Adult help is required for this experiment; copper sulfate is poisonous if swallowed. Electroplating can be a simple and effective process, as you can see by plating a metal object such as a brass key with copper. Make up a solution of copper sulfate (available from most pharmacies) using two (plastic) teaspoons of powder in distilled water. Take a lead from the negative terminal of the battery to the key. Strip the insulation from 3 in (8 cm) or so of wire, twist into a coil, and connect to the positive terminal of the battery. Twist the bare wires into a coil; then dip this and the key into the solution.

YOU WILL NEED
● *container* ● *battery* ● *two leads* ● *short length of copper wire* ● *metal object to electroplate (such as an old brass key)*

What happens
After a few minutes, you should see the key begin to get a coating of copper. The electric current has split the copper sulfate into its components, and the free copper has been drawn to the negative charge on the key. Copper from the coil keeps the process going.

Before   *After*

EXPERIMENT
Electrolysis

Adult supervision is required for this experiment. This experiment shows how water is split into its two components, hydrogen and oxygen, by electrolysis. When an electric current is sent through water, bubbles of gas form around the two electrodes. Gas from the anode relights a glowing taper, showing that it is oxygen. Gas from the cathode ignites with a "pop," showing that it is hydrogen.

YOU WILL NEED

● *large tank or basin* ● *2 yogurt containers* ● *2 test tubes, or narrow glass jars (we have used wide jars only for clarity)* ● *2 graphite pencils or the carbon rods from the center of two old batteries* ● *salt (or sodium carbonate - washing soda)* ● *two leads* ● *battery (9V)*

1 MAKE A SMALL HOLE in the base of each yogurt container to hold the graphite pencils.

2 ATTACH A LEAD to each pencil inside the container. Cut a notch in the rims of the containers so that they can rest flat on the bottom of the bowl.

Electrolysis in action
Attach one lead to each of the battery terminals. Soon gas will start to bubble up from the pencils in each jar and the water level will start to drop.

3 PLACE THE container open end down in the bowl, with the leads hanging outside. Fill the bowl carefully with tepid water in which the salt, or sodium carbonate, has been completely dissolved. Avoid disturbing the containers.

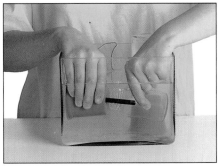

4 PUT EACH JAR in the bowl and fill with water. Gently slot a pencil into each jar and stand it upright, keeping the jar's neck underwater at all times.

Hydrogen and oxygen
Because water has two hydrogen atoms for every oxygen atom, twice as much hydrogen forms.

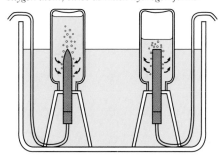

Oxygen collecting in the top of the jar with the pencil attached to the battery's positive terminal.

Hydrogen collecting in the top of the jar with the pencil attached to the battery's negative terminal.

Add half a teaspoonful of sodium carbonate to the water. As this dissolves, it makes the water a better electrolyte, which means that electric current can pass through it easily.

To test the gases, lift the jars from the water. Keep them upside down and away from you. Hold a lighted taper under the negative terminal jar, and a glowing taper under the positive terminal jar.

MAGNETISM

Magnetism is often thought of as just the invisible force that makes compasses point north and makes magnets attract and repel each other. But it is also closely related to electricity. This relationship, known as "electromagnetism," is one of the most important in physics. Every kind of radiation, from radio waves to visible light, is electromagnetic; all but a tiny fraction of the world's electricity is generated electromagnetically; and an enormous range of devices, from doorbells to television, work electromagnetically.

The spectacular aurora *of colors often seen in the night sky over the poles is the glow of ions (p. 156) in the air energized by rays from the Sun. The Earth's magnetic field draws them over the poles.*

Magnetic effects were first discovered long ago in the black rock lodestone, found all over the world. Lodestones were used as compasses by sailors more than a thousand years ago.

People have probably known since prehistoric times that lumps of a certain type of rock can attract or repel each other, depending on which way they are turned. This rock is called lodestone, and the iron oxide in it makes it naturally magnetic.

The Ancient Greeks knew of lodestones, and magnets get their name from the Ancient Greek town of Magnesia, where many such stones were found. Greek legends tell of magnetic mountains that drew the iron nails from ships or stuck a shepherd's feet to the ground with the iron tacks in his shoes. According to tradition, the Greek philosopher Thales of Miletus (624–546 B.C.) was the first to describe this phenomenon, in about 550 B.C.

Yet the Ancient Chinese knew about magnets long before the Greeks. They knew, for example, that magnets point north if allowed to swing freely. As early as 2500 B.C., a Chinese emperor is said to have used a lodestone to guide his troops through fog. The Chinese are also believed to have discovered that an iron needle can be magnetized by heating it up and letting it cool while aligned in a north-south direction.

No one knows who first thought of putting a magnetized needle on a pivot and setting it in a box to make a compass. Legend has it that the Chinese passed the idea on to the Arabs, who, in turn, passed it on to the Europeans. Sailors were probably using compasses to help them find their way long before 1269, when the earliest description of a compass was written by the French scholar Peter Peregrinus (c.1240–unknown).

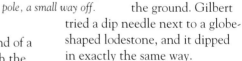

Magnetic compasses point not to the true North Pole, but to the Earth's magnetic pole, a small way off.

■ Poles apart

Peregrinus named the end of a magnet that points north the North Pole, and the other end the South Pole, and showed that like poles repel each other and unlike poles attract. He also described a now-famous way of showing that magnets create an area of attraction and repulsion around themselves by shaking iron filings onto a piece of paper above a magnet.

Yet though Peregrinus and his contemporaries knew that magnets pointed north, no one knew quite why. Some thought there was a vast lodestone mountain in the far north to which the compass needle was attracted. The problem was not solved until

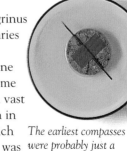

The earliest compasses were probably just a magnetized needle floated in a bowl (p. 162).

1600, in the reign of Queen Elizabeth I of England. That year, William Gilbert, personal physician to the Queen, published a book called *De Magnete*, recounting the results of his experiments with magnets.

Dr. Gilbert showed that the Earth itself is a giant magnet. He mounted a magnetized needle so that it could pivot freely in all directions, up and down and side to side. This is called a "dip" needle, because the north pole of the needle dips down toward the ground. Gilbert tried a dip needle next to a globe-shaped lodestone, and it dipped in exactly the same way.

For a long time, scientists believed the Earth must have a giant iron magnet as its core. Yet although the Earth's core is iron, this cannot be a magnet. It is now known that iron loses its magnetic properties above 760°C (called the "Curie temperature," after the French scientist who discovered it in 1895, Pierre Curie). Yet the Earth's core is at least 1000°C. The real key to the Earth's magnetic properties seems to lie in the link between magnetism and electricity, a link first discovered early in the 19th century.

The first inkling of a

link came in 1819, when Danish physics professor Hans Christian Oersted (1777–1851) was giving a lecture on electricity. To show how an electric current heated wire, he connected the wire to the terminals of a big battery. To his surprise, the needle of a compass sitting on the bench near the wire at once swung around — only to swing back again to its normal position pointing north when the wire was disconnected. The possibility of a link between electricity and magnetism had long been suspected, and Oersted realized that an electric current has magnetic effects just like a lodestone or a steel magnet.

The strength of the magnetic movement provides a simple way of measuring currents in electric meters.

■ Electromagnetism

Soon dozens of scientists were conducting their own investigations into this exciting phenomenon. Months after Oersted announced his discovery in 1820, French physicist André Ampère (1775–1836) proved that two parallel wires carrying currents in the same direction attract each other, just as unlike poles of a bar magnet, while wires with currents flowing in opposite directions repel each other. Ampère also showed that if you made a cylindrical coil of wire (now called a solenoid), it behaved just like a bar magnet.

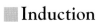

Doorbells use the magnetism of electrified coils to pull a hammer onto a bell (p. 165).

By 1825, English physicist William Sturgeon found that the strength of an electromagnet could be boosted dramatically by placing a bar of soft iron inside the coil of wire. American physicist Joseph Henry (1797–1878) improved on this design by insulating the wires in the coil. In 1831, Henry supervised the making of an electromagnet capable of lifting over a ton.

It was the remarkable English scientist Michael Faraday (1791–1867), however, who made the most dramatic contribution to the the budding science of electromagnetism. Son of a poor blacksmith, Faraday grew up to become one of the greatest experimenters of all time.

First of all, he repeated Peregrinus's experiment with iron filings, with both bar magnets and electric coils, and suggested that the lines of filings marked real lines of electromagnetic force. He called the area around a magnet in which these lines of force occur its "field," and the concept of field has become one of the most important in physics.

■ Induction

In the 1830s, Faraday began to delve deeper into the relationship between electricity and magnetism. If a current can create a magnetic field, he asked, can a magnet create a current? To test this possibility, he set up two coils of wire next to each other, one with an iron bar inside to create a strong magnet. When he sent a current through the coil with the iron core, Faraday thought its magnetic field might create a current in the second coil as long as the magnet was switched on. Instead, there was just a brief surge of current in the second coil and then nothing.

The reason, as Faraday quickly realized, is that it is not a magnetic field that creates a current in a wire, but the *movement* of magnetic lines of force across the wire. He proved this by moving the iron bar in and out of the coil, and by moving a loop of wire across a magnetic field. Each time, a current was "induced" in the wire. Faraday's discovery of "electromagnetic induction" — a discovery made independently by Joseph Henry around the same time — may have had more impact on our world than almost any other single scientific discovery. Using this principle, machines could be built to generate huge quantities of electricity, opening the way for everything from electric lighting to telecommunications.

■ Waves

Two years before Faraday died in 1867, Scottish physicist James Clerk Maxwell made a crucial breakthrough in the understanding of electromagnetism. Wheras Faraday was a great practical scientist, Maxwell was a brilliant theorist. Developing Faraday's idea of fields, he reduced everything known about electricity and magnetism to just four basic mathematical equations — all of which worked only if electromagnetic fields rippled through space as waves. Since, according to his calculations, these waves must move at a speed very close to the speed of light, he rightly decided that light was just one of many forms of electromagnetic wave (pp. 78-79).

On the magnetic levitation railroads, the power of electromagnetic repulsion supports the entire weight of a train invisibly a few inches above the ground. This makes for a smooth, almost frictionless ride. The same force is used for the "linear induction motors" that draw the train along at high speeds.

Sailors have long relied on magnetic compasses to help them navigate. Today, ships use many other electromagnetic effects, including radar. Radar works by plotting the pattern of echoes of pulses of electromagnetic radio waves.

Poles and fields

YOU CANNOT SEE MAGNETISM, but you can feel its effects. It is an invisible force that acts on iron, steel, and a few other metals, and often either pulls them together or pushes them apart. Try to push two strong bar-shaped magnets together one way, and they will fight against it. Turn one around the opposite way, and they will snap together by themselves as soon as they are close enough to feel each other's influence.

There is a definite area around every magnet where the magnetic force exerts its power — the magnetic "field." This field gets gradually weaker farther away from the magnet. But there are two areas where it is especially strong. These are the magnetic "poles." They are called poles because the Earth itself is a giant magnet, and all magnets are influenced by its field. When you hang a magnet from a thread so that it can rotate freely, it always ends up pointing the same way, with one end pointing to the North pole and one to the South. This is why the two ends of a magnet are called the North (or North-seeking) pole and the South (or South-seeking) pole. All magnets have these poles, whether they are long and bar-shaped, square, round, or horseshoe-shaped.

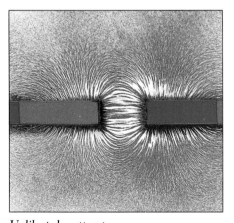

Hanging together
The effects of magnetic attraction are transmitted between objects affected by magnetism. So a magnet can pick up a whole string of steel ball bearings.

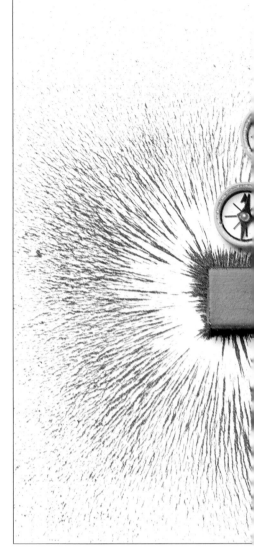

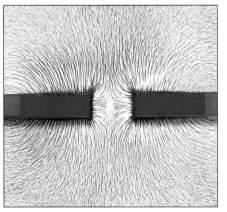

Unlike poles attract
When magnets are put together, N pole to S or S to N, the field between the two adjoining poles is strengthened. Lines of force run in straight lines between them, and the field curves almost as if there were another magnet there.

Like poles repel
When magnets are put together, N pole to N or S to S, the field between the two adjoining poles is weakened so much that a neutral spot appears between them. All the lines of force curve away from this spot.

■ Attraction and repulsion

Put two magnets together, and they will either snap together or spring apart. If the North pole of one magnet meets the South pole of the other, or South meets North, the magnets will snap together. But if North meets North or South meets South, they spring apart. This is because like poles always repel each other and unlike poles always attract. Just what happens to the magnetic field when you put two magnets together can be shown with iron filings in the same way as the field for a single magnet above. Notice how the lines of force shown by the filings never cross or intersect, for they simply show which way a magnet points at any one place.

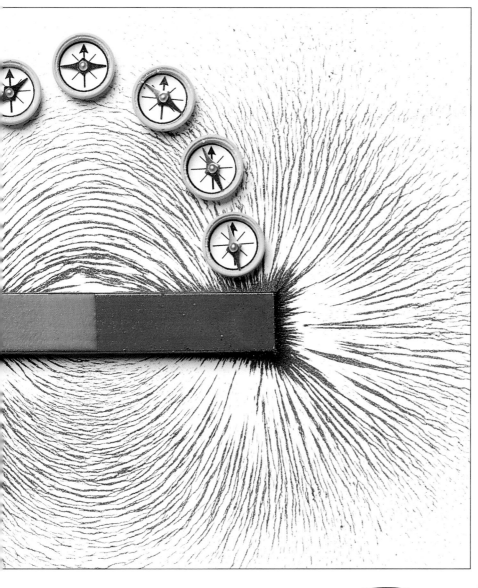

Magnetic field

The lines of magnetic force around a magnet — the field — can be shown very clearly with small filings of iron and tiny compasses. Whenever an unmagnetized magnetic material such as iron comes near a magnet, it becomes temporarily magnetized — that is, it has North and South poles, which are drawn toward the opposite poles of the magnet. So if iron filings are scattered around a magnet, they instantly swivel to align with the magnet's field. They cluster around the poles, where the field is strongest, and only friction prevents filings farther away from the poles from sliding toward the magnet. The needles in "plotting" compasses are magnets already, but they show the field in the same way.

Force field

Iron filings and plotting compasses strikingly reveal the lines of magnetic force around a magnet. Note the clusters of filings around the poles (ends) of the magnet. The field is shown here in one plane only, but it extends in a similar pattern in all directions around the magnet.

Magnet Earth

The Earth is a giant magnet — which is why magnets always point in the same direction if allowed to swivel freely. In fact, its field extends some 50,000 miles out into space, creating such weird effects as the aurora, the spectacular colored lights that sometimes appear in the night sky over the poles. Yet no one really knows why Earth is a magnet. After all, its core is not solid iron, but liquid. One explanation is that electric currents circulate within the core, turning it into a giant solenoid (p. 164). Whatever the cause, it is known that the Earth's North Pole and South Pole have switched over at least twice since the Earth was formed.

Magnetic barrier

It's as if there were a giant bar magnet through the middle of the Earth from the North to South pole — the magnetic North pole, though, is in the sea north of Canada, some way from the world's true North Pole. The lines of force extend far out into space, forming a barrier around the Earth called the "magnetosphere."

161

Magnets

CUT A MAGNET IN HALF, and you get two new magnets, each with North and South poles. Cut it in half again, and you get two more magnets. No matter how often you divide a magnet, up to a point, you always get new magnets. No one knows why, but it is thought that all magnetic materials are made up of tiny groups of atoms called "domains," each of which is like a mini-magnet with its own N and S pole. In any iron or steel bar, there are many of these domains. When the bar is unmagnetized, their poles point randomly in any direction. But when you put a magnet nearby, they all begin to point the same way. In steel, once they are lined up, they tend to stay that way, and the steel becomes a permanent magnet. It only loses its magnetism if the domains are jumbled up again — perhaps by hitting with a hammer or heating to red heat. Iron, however, loses its magnetism almost instantly.

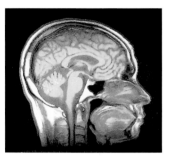

Brain scan
Strong magnetic fields enable doctors to see inside the body using nuclear magnetic resonance (NMR) scans.

All-around field
Some magnets are mounted so that they can swivel in all directions. With these you can see that a magnet's field extends in all directions.

EXPERIMENT
Making a compass

You can turn a steel needle into a magnet very simply by stroking it in the right way with another magnet. Once it is magnetized, you can use the needle as a pointer for a simple home-made compass that will always show you where North and South are.

YOU WILL NEED
● *saucer* ● *slice of cork* ● *needle or metal rod* ● *tape*

1 TO MAGNETIZE the needle, draw a magnet along it repeatedly in the same direction for 15 seconds or so.

2 TAPE THE NEEDLE to a broad, flat cork, and float the cork in a saucer of water. The North pole of the needle will always turn to the North.

EXPERIMENT
Magnetic boats

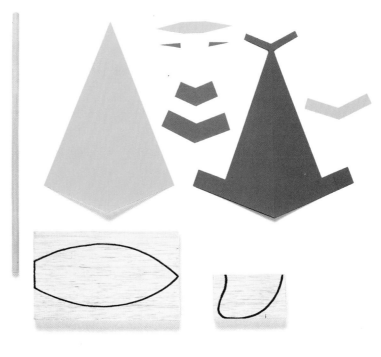

Adult supervision is required.
Magnetism can be felt through many materials, even through glass.

YOU WILL NEED
- *thin dowel* ● *colored paper* ● *paper and wood glue*
- *balsa wood* ● *corks* ● *pins* ● *screws* ● *bar magnets*
- *knife* ● *scissors*

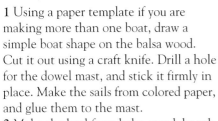

1 Using a paper template if you are making more than one boat, draw a simple boat shape on the balsa wood. Cut it out using a craft knife. Drill a hole for the dowel mast, and stick it firmly in place. Make the sails from colored paper, and glue them to the mast.

2 Make the keel from balsa wood; knock three nails carefully into the bottom, and stick it to the underside of the hull.

3 Fill a large bowl or fish tank with enough water for the boat's keel to float just above the bottom.

4 Make up marker buoys with corks weighted with brass screws.

5 Hold a magnet under the tank, and use it to draw around the tank between the buoys. If the magnet sinks the boat, hold it farther from the tank.

Boat race
For clarity, only one large boat and two buoys are shown here in a small tank. But the game is much more fun if you make a number of smaller boats or use a much larger tank. Set up an elaborate zigzag course with many buoys. Then you can have races with your friends, either all at once or against the clock.

The electric magnet

PUT A SMALL COMPASS near a wire carrying an electric current, and a surprising thing happens. The compass needle swings around as if next to a strong magnet. In fact, this is just what is happening; an electric current creates its own magnetic field similar to that around a steel magnet. Indeed, electricity can be used to create strong magnets, called "electromagnets," that can be switched on and off as easily as an electric current. If the wire is formed into a loop, the magnetic effect is much stronger. It is even stronger if the wire is wound into many loops. So most electromagnets are made as coils of wire, called "solenoids." A rod of iron down the middle further boosts the power of the electromagnet.

Pulling power
This electromagnetic crane in an auto scrapyard provides convincing demonstration of the power of electromagnetism.

EXPERIMENT
Making an electromagnet

Adult supervision is suggested. Make a simple electromagnet with enameled copper wire (from electrical shops) and an iron rod. An iron nail will suffice if you cannot find a rod.

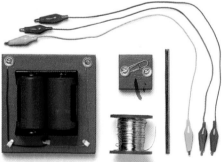

YOU WILL NEED
● *battery* ● *switch* ● *three leads* ● *iron rod*
● *enameled copper wire*

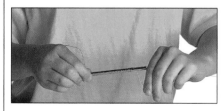

Making a solenoid
Wrap the copper wire tightly around the rod, keeping the coils as close as possible, using adhesive tape to hold the coils in place. Then connect the loose ends of the wire to the batteries with a simple switch.

Instant lift
Switch on the current, and the solenoid will be magnetized instantly. Test it by picking up paperclips and nails. When you switch it off, the magnetism will disappear rapidly.

EXPERIMENT
Making a buzzer

Adult supervision is required.
Use the power of electromagnetism to create this electric buzzer. With the current on, the solenoids are magnetized, pulling the spring to hit the hammer. But as the spring moves, it breaks the circuit, the magnets switch off, and at once the same sequence begins again.

YOU WILL NEED
● *wooden base* ● *clean food can* ● *dowel* ●
balsa block ● *thumbtack* ● *nails* ● *circles
of card* ● *2 screws and eyelets* ● *enameled
copper wire* ● *3 leads* ● *switch* ● *battery*

1 GLUE THE DOWEL upright on the wood base. Make sure there are no sharp edges on the empty food can, and put a nail through the center on the dowel.

2 TAKING GREAT CARE to avoid cutting fingers on sharp edges, use clippers to cut a rectangle from a soda can. Then cut out the spring as shown below.

3 BEND THE SPRING as shown in the diagram, and glue a 4 in (10 cm) nail inside to make the hammer. Then screw the flap of the spring to the wooden base.

4 USING A HACKSAW, cut two rods, each about 1 in (2.5 cm) long, from a long nail. Stick circles of card on each end, and wind the wire tightly around each.

5 MAKE SURE you wind the wire onto *both* nails in one long strand to make the solenoids. Glue them down so that the nails do not quite touch the spring.

6 WRAP WIRE from one solenoid around a thumbtack, and push into the balsa block. Glue the block on the board so that the tack just touches the spring.

and switch as shown. When you switch on, the spring should whir to and fro, hitting the nail on the food can.

How to wind the wire onto the nails to make the solenoids

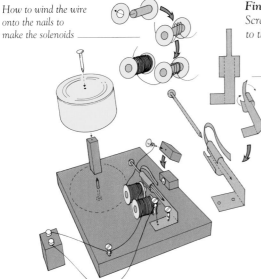

Final connections
Screw down the free wire from the solenoids to the board. Then connect up the battery

The spring and hammer are shown in gray

If the buzzer does not work...
adjust the gap between the spring and the solenoids and the spring and the thumbtack.

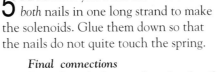

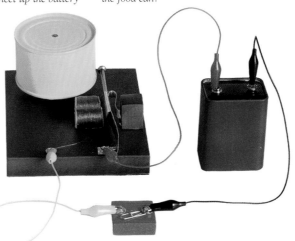

Generating electricity

JUST AS ELECTRIC CURRENTS create their own magnetic fields, so magnets too can create electric currents. Move a magnet near a coil of wire, and a current is generated in the wire. Move the wire near the magnet, and the same happens. It makes no difference which moves, wire or magnet; it is the movement of the wire in relation to the magnetic field that creates the current. The field is said to "induce" a current in the wire, and the effect is called "electromagnetic induction."

Generators exploit this effect to provide over 99 per cent of our electricity. Most work by spinning magnets between coils of wire — the stronger the magnet, the faster it turns and the more coils there are, the bigger the voltage created. Dynamos give a "direct current" that always flows in the same direction. Alternators give an "alternating current" that continually switches in direction — because as the magnets spin, they pass the wires going upward on one side and downward on the other. But the principle is the same.

Power source
Inside the generators in most power stations are huge magnets that rotate inside a ring of electric coils. In coal, oil, gas, and nuclear power stations, the fuel makes steam to turn turbines and drive the magnets around. In hydroelectric stations (p. 134), the turbines are driven by moving water.

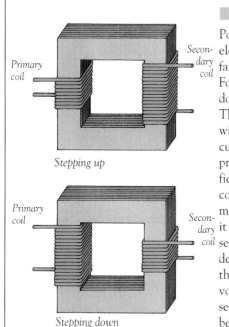

Primary coil

Secondary coil

Stepping up

Primary coil

Secondary coil

Stepping down

■ Transforming current

Power station generators push out electricity at 25,000 volts or more — far too powerful to use in the home. Fortunately, voltages can be stepped down (or stepped up) with transformers. These have a soft iron core wrapped with two coils of wire. If an alternating current is passed through one coil (the primary coil), it creates a magnetic field in the core that changes direction continually with the current. As the magnetic field switches back and forth, it induces a new voltage in the other, secondary coil. The size of this voltage depends on how many turns of wire there are in each coil. To step up the voltage, there must be more coils in the secondary; to step it down, there must be fewer.

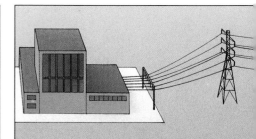

■ Power to the home

Electricity from power stations is fed into a network of cables — a grid. Transformers boost the voltage from the generators up to 400,000 volts, because at high voltage, less energy is lost through resistance in the cables. Along the way, it is stepped back down to provide voltages for industry, factories, and homes, where it is typically 110–240 volts, depending on the country.

EXPERIMENT
Inducing a current

This experiment shows how you can generate an electric current by moving a wire through a magnetic field. Remember that the wire must be part of a complete circuit for the current to flow. If you connect the wire to a sensitive current meter (most home ammeters or multimeters will do), you should be able to detect the current generated.

YOU WILL NEED
● two bar magnets ● length of copper wire
● two leads ● meter for measuring current (ammeter or multimeter)

Adult supervision is suggested.

■ Fleming's right-hand rule

When you move a wire through a magnetic field, the current created always flows the same way — at right angles both to the lines of force in the field and to the movement. Fleming's "right-hand rule" is an easy way of remembering which way the current will flow.

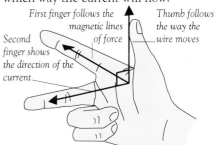

First finger follows the magnetic lines of force
Second finger shows the direction of the current
Thumb follows the way the wire moves

Hold your right hand as shown with the thumb and first two fingers at right angles to each other.

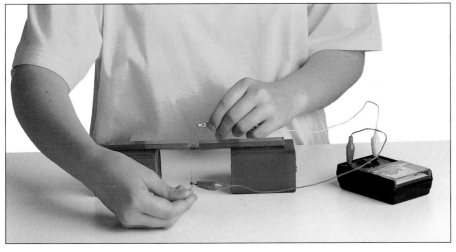

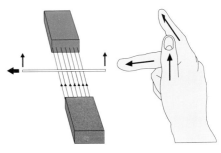

Cutting the field
Set up the magnets with opposite poles facing. Move the wire quickly up or down through the gap. You should see a tiny deflection of the meter needle; moving the wire faster deflects the needle more, indicating a stronger current.

To use the right-hand rule, point your first finger in the direction of the magnetic field, and your thumb the way the wire moves. Your second finger will show the induced current.

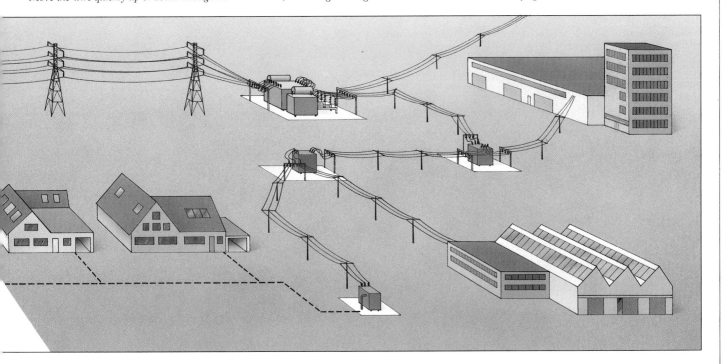

Personal telephone

THERE ARE MORE than 500 million telephones in the world today, transmitting almost instantly not only voices but also words, pictures, computer data, and even videos. Sometimes the message travels just a few meters along a cable; sometimes it is beamed by satellite all the way around the globe. But wherever they go, all telephone calls work the same way. First, the message — whether a voice or a picture — is converted into a varying electric current using electromagnetism. Then, at the other end of the line, it is converted back the same way to re-create the sound of your voice or the details of the picture. In the mouthpiece of most phones, there is a thin disc called a diaphragm. This vibrates when you speak, disturbing a magnetic field and altering an electric current. At the receiving end, the changes in the current switch an electromagnet on and off, vibrating another diaphragm and reproducing the sound. The homemade telephone shown here works in much the same way.

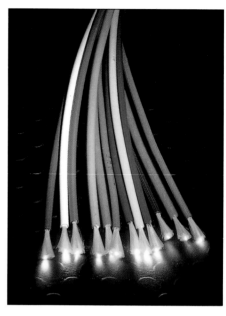

Light conversation
Fiberoptics are now taking over from traditional copper telephone cables. They transmit signals very rapidly as flashes of light reflected from side to side through thin strands of glass, and they can carry far more messages than copper cables.

EXPERIMENT
Making a telephone

Adult supervision is required.
Use the sound of your voice to generate a tiny electric current to transmit a message.

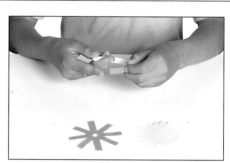

1 TO MAKE the diaphragm, cut from thin card two strips and a star as shown in orange (left). Bend one strip into a ring the same diameter as the pot near the top. Cut a paper circle as at left in white.

4 WIND ENAMELED COPPER WIRE tightly many times around the tube, leaving both ends free. This is your solenoid. Glue one end of the the solenoid firmly into the center of the diaphragm cone.

■DISCOVERY■
The first telephones

Bell's second telephone worked in much the same way as the simple telephone made on this page. The caller shouted into one cone, vibrating a magnet and creating a current. This set a diaphragm in the receiver vibrating.

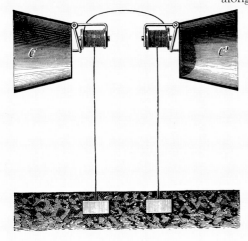

THE INVENTION of the telephone in 1876 is credited to Alexander Graham Bell, although others were working on similar devices. Electric telegraphs were already used to send messages along a cable by switching the current on and off. Bell found a way of carrying the vibrations of the voice in a similar electric signal. He placed an iron disc in front of a solenoid so that when someone spoke, the disc vibrated, inducing a current in the coil and sending electric pulses through the circuit. One day in 1876, Bell was working upstairs on his invention. Suddenly, his assistant heard Bell's voice on a receiver downstairs, "Mr. Watson, come here, I want you": the first words ever spoken over a telephone.

You may have to shout to generate a strong enough current!

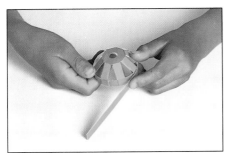

2 SCORE THE STAR SHAPE along the line marked, and bend the spokes up to form a shallow cone. Stick the ends of the spokes to the outside of the ring. Wrap the other strip around, and stick.

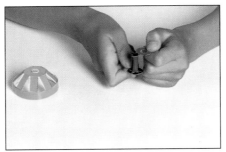

3 STICK AN IDENTICAL cone inside the first. Cut out the blue concertina shapes (left). Bend around a pencil, and glue to make a tube. Bend over the tabs at each end, and stick on two card rings.

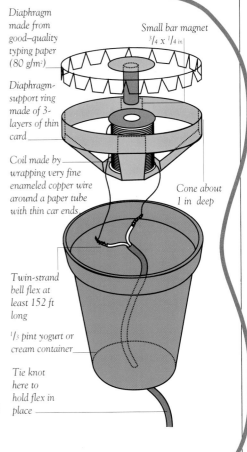

Diaphragm made from good-quality typing paper (80 g/m²)

Small bar magnet ³/4 x ¹/4 in

Diaphragm-support ring made of 3-layers of thin card

Coil made by wrapping very fine enameled copper wire around a paper tube with thin car ends

Cone about 1 in deep

Twin-strand bell flex at least 152 ft long

¹/3 pint yogurt or cream container

Tie knot here to hold flex in place

5 GLUE A SHORT BAR MAGNET (20 mm by 5 mm) to the center of the white disc at right angles. Slot the rod into the coil (make sure it does not touch the sides), and glue the disc to the cone.

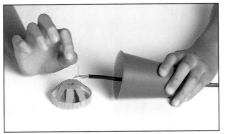

6 SLOT A LONG two-stranded cable into the container, and connect one strand to each end of the coil. Push the diaphragm into the container. Now repeat all steps to make the other phone.

Electric motors

GENERATORS use magnets to turn motion into electricity; electric motors use them to turn electricity into motion. Motors work because when a current flows across a magnetic field, the wire recoils. If the wire is in a loop, current flows out one side and back the other — so the reaction between current and field pushes the loop down one side and up the other, making it spin. To keep the loop spinning the same way, the simplest motors need an alternating current to continually reverse the current, since first one side of the loop and then the other pass each pole of the magnet. With direct current, a device called a "commutator" continually reverses the direction of the current.

Fast train
France's TGV, one of the world's fastest trains, is powered by electric motors. These are not only clean and quiet; they can also provide very quick acceleration because the moving parts are light compared to diesel and steam engines, and there is no fuel to carry. Electric current is picked up from overhead cables through flexible connectors called pantographs.

EXPERIMENT
Making a current meter

The movement generated by the reaction between magnets and an electric current is called "electromotive force," and you can use it to make a simple current meter. In most motors, the magnets stay still, and it is the wires that move. In this meter, however, the magnet — the needle of a compass — moves and the wires stay still. The farther the needle moves, the stronger the current.

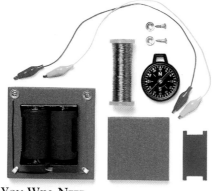

YOU WILL NEED
● *batteries* ● *square balsa base* ● *pocket compass* ● *bare copper wire* ● *strip of card* ● *two brass screws and eyelets* ● *two leads*

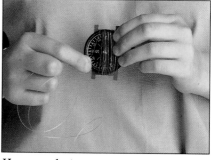

How to make it
Stick stiff card (shown in blue) to back of the compass, and wind the copper wire around both about 20 times. Mount on a block of balsa, and connect the leads. The movement of the compass needle shows the current.

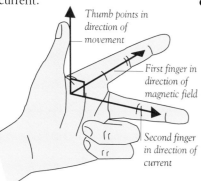

Thumb points in direction of movement

First finger in direction of magnetic field

Second finger in direction of current

Fleming's left-hand rule
Like the right-hand rule for generators, the left-hand rule gives the direction of movement for electric motors.

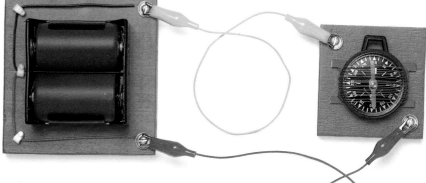

 Adult supervision is suggested.

EXPERIMENT
Making a motor

 Adult supervision is suggested.

Trains use huge motors with coils of special copper thousands of turns long, but you can make a simple electric motor with ordinary copper wire and basic household items. You will need at least a 4.5-volt battery, and although your motor is not powerful, you could cut out paper fan blades and make your own fan.

YOU WILL NEED
- square wood base • three split-pin clips
- two metal right-angled brackets with screws • two strong magnets • thin copper tubing • metal rod • balsa block • enameled copper wire • insulating tape • single-core wire • four screws • two brass screws and eyelets • three leads • switch • battery

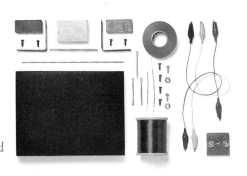

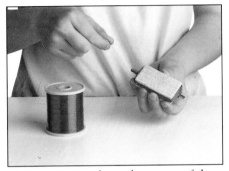

1 DRILL A HOLE down the center of the balsa block, and thread a copper rod through. Cut grooves down each side of the block, and wind copper wire very tightly around it.

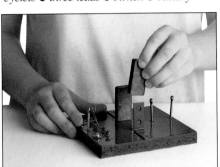

2 SCREW THE BRACKETS to the base, and stick the magnets to them. Drill holes for the eyelets in a line down the middle. Make screw terminals, and bend the wire, as shown below, to make the contacts.

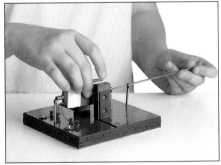

3 MAKE THE COMMUTATOR in foil as described below, slot the rod through the eyes and the coil, and connect the terminals to a battery. You might need a gentle push to get the coil spinning.

Cut grooves down each side of the balsa block

Making the commutator
The current gets to the spindle (A) of the rotating coil from the fixed wires (B and C) via the commutator (D). This reverses the current every half-turn to keep the motor spinning. To make it, wrap tape around the spindle, then tape the wires from the coil along this tape down opposite sides of the spindle. Bend B and C so that they cup around the spindle – – they should contact the wires on the spindle but not each other.

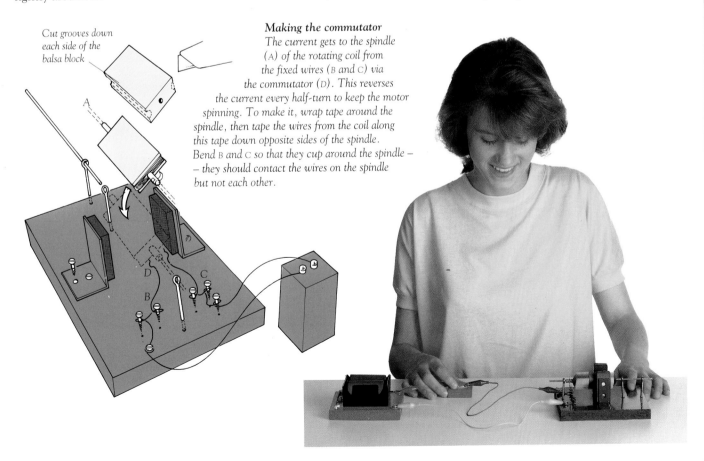

Electronics
AND
computing

ELECTRONICS IS ONE of the newest of all sciences, barely half a century old. Yet already, it would be hard to imagine a world without electronic devices such as radios, televisions, and computers. Electronic technology enables us to do anything from flying fast and safely to communicating instantly with the opposite side of the world.

Jet airliners are coming to rely more and more on electronic technology, as the electronic displays on the flight deck testify. Already the state of the aircraft is monitored electronically, and the plane is navigated and landed with the aid of electronic equipment. In future, many planes will be "fly-by-wire", which means that the control flaps are operated not mechanically but electronically.

ELECTRONICS

FEW BRANCHES OF SCIENCE have had more impact on our lives than electronics. By using tiny electric currents to switch electric circuits on and off, it is possible to make electronic systems operate very complicated processes automatically. Thanks to recent advances in electronics, it is also possible to build huge computing power and very sophisticated control systems into tiny devices, and electronics are now used for an increasing range of tasks, from automatic exposure and focus in a camera to navigating an airplane.

Computer technology has developed rapidly with the invention of the microchip, and even inexpensive desktop models are now able to create complex graphics. This picture shows a computer being used for engineering design work.

A basic radio receiver can be built with a transistor and just a few other electronic components (p. 178).

The origins of electronic technology date back to the second half of the 19th century, when scientists were first beginning to understand the nature of electromagnetic radiation.

At that time, the German physicist Heinrich Hertz (1857–94) was looking into the effects of big sparks in the air. In 1888, he built a circuit to create regular high-voltage sparks across a gap between two metal balls. If Maxwell was right (p. 159), the sparks would send out waves of electromagnetic radiation. So Hertz set up another circuit to detect the waves.

■ Radio waves

Just as expected, the waves created small pulses of current in the receiving circuit, which Hertz saw as tiny sparks across another gap. By moving the receiver around the room, he could work out exactly what size these waves were. They proved to be much longer than waves of light (pp. 78-79), and are now known as radio waves.

Scientists in the 1890s were fascinated by the idea of using such waves to transmit messages from one place to another. But they needed a receiving circuit sensitive enough to pick up the waves over anything more than a few yards. In 1890, the French physicist Édouard Branly invented a sensitive device called a coherer — basically a glass tube full of iron filings between two electrodes — which worked over about 500 ft (about 150 m).

■ Marconi

Five years later, the young Italian inventor Guglielmo Marconi (1874–1937) managed to pick up signals 1 mile (1.6 km) across his family's estate. Eight years later, he transmitted a message in Morse code — that is, in coded electrical pulses — over the English Channel. And in 1901, Marconi became a household name by beaming a signal right across the Atlantic, from Cornwall in England to a wire or antenna dangling from a kite high above Newfoundland. Radio was born.

By this time, receiving circuits could be tuned to pick up particular wavelengths and so respond to signals sent out from different transmitters at different

Resistors are a vital part of any electronic circuit, protecting the transistor and controlling the current.

wavelengths. The American physicist Reginald Fessenden developed a way of modulating (varying) radio waves to mimic sound waves. This meant that speech and music could be transmitted, not just Morse code. Soon thousands of people were tuning in to radio broadcasts. Their receivers contained crystals that allowed a current to pass only one way, for modulation meant that the radio signal had to be generated by an alternating current, not a spark.

■ Valves

Unfortunately, the signal was so weak that listeners had to hunch over their crystal sets with headphones over their ears. For a reasonably loud sound, the signal would have to be amplified. The answer was found in an effect noticed by the American inventor Thomas Edison (1847–1931) in 1883, when he was trying to create lightbulbs (p. 150). In one experiment, he sealed a metal wire inside the bulb along with the filament, and, to his surprise, a small current flowed from the hot filament to the wire — but not the other way round.

In 1904, Ambrose Fleming (1849–1945) used this effect to

create the first genuine electronic component, the thermionic valve or diode (p. 176) – – a glass tube with two electrodes: a heated filament and a metal plate. Since the valve allowed current to pass only one way, it rectified (converted) the alternating current created by the radio signal into a simple direct current. Two years later, American scientist Lee de Forest discovered how to make valves amplify. He found that he could increase the current flowing between the electrodes by placing a third electrode (a metal grid) between the filament and the plate. Moreover, he could change the increase at will by varying the small current in the grid.

The microchip was originally developed for space technology, and the U.S. space shuttle now relies on complex electronic circuits for controlling every-thing from the craft's course to its air supply.

▪ Triodes

De Forest's three-electrode valve, or "triode," was the basis for the development not only of radio, but of all kinds of electronics. Television, for example, would not have been possible without the triode. Like radio, television uses electromagnetic waves to transmit the signal, and the triode was essential to amplify this signal. TV cameras rely on the photoelectric effect (p. 13) to turn pictures into an electrical signal. In the receiver, the TV signal is used to re-create the picture in a cathode ray tube.

Triode valves were the basis of all electronic equipment until the 1940s. Even early computers used valves. A simple computer could contain as many as 20,000

Advances in electronics have spurred the development of interactive video/computer systems. This is an IAV disc.

valves, and fill an entire room. The valves generated so much heat that frantic engineers had to work both day and night to replace overheated valves.

The real breakthrough in electronics came with the development of transistors in the 1940s. Transistors are based on special materials called semi-conductors, such as germanium and silicon, which conduct electricity only partially. Scientists at the Bell Telephone Laboratories in the U.S. found that by implanting electrodes in such solid lumps of these materials, they could create a tiny device that worked exactly like a diode valve.

Transistors come in various shapes, but they are all triodes, with three terminals.

▪ The transistor revolution

In 1948, three scientists at the Bell Laboratories — William Shockley, Walter Brattain, and John Bardeen — made a semi-conductor device that worked like a triode. This device came to be known as a transistor and revolutionized electronics. Tiny transistors rapidly replaced valves in

radios, television, and computers, making them all much smaller, more reliable, and robust. By the 1960s, valves were largely a thing of the past, and radios were essentially transistor radios.

With the aid of the transistor, computers developed dramatically in the 1950s, but they were still very small and very expensive. Then, in 1958, there was another breakthrough in electronics. This was the invention of the microchip by American Jack Kilby. Kilby put the connections for two transistors inside a single crystal of silicon, just $^3/_8$ in long. This was the first "integrated circuit," because all the circuitry was packed into this one crystal.

▪ Microchip technology

Soon chips were becoming smaller and smaller and electronic circuits were becoming more complicated, as scientists discovered different ways of squeezing more and more components into a single chip. Today, silicon chips range from simple circuits for electric toasters to microprocessors containing a million or more transistors.

The invention of the microchip meant that highly complicated electronic circuits could be packed into a tiny space and reproduced very cheaply. They are now found everywhere, not only in computers, but in pocket calculators, cameras, watches, personal stereos, microwave ovens, and much more besides. But the science is now developing at such a pace that no one can predict what the future will bring.

Besides transistors and resistors, most electronic circuits also include capacitors. Capacitors are, in a way, like rechargeable batteries. They can store a small charge which can be released when needed to make a current flow for a short while — to light a bulb, for instance. When a diode is used to rectify an alternating current, the current continually stops, because the diode allows current to flow only one way. A capacitor can release its charge to fill in the gaps.

In future, the automobile is likely to have more and more electronic functions. Things like ignition and fuel supply are already controlled by microchips in many cars. Soon electronic displays may replace the more familiar instrument panel inside the car, and the driver may be guided through cities by a computer continually fed with the latest traffic information.

Electronic circuits

ELECTRICITY IS A CLEAN and convenient source of power –
– but it can also be used to send signals carrying
information. Television, radios, computers, and other
"electronic" devices use electricity this way. Inside every
electronic appliance, circuits carrying tiny electric
currents are continually switched on and off or varied in
strength to send signals that tell the appliance what to
do. In a computer, for example, tiny currents switch on
and off to represent numbers. In a stereo system, the
currents control the movement of the loudspeaker
diaphragm so that it reproduces the right sound.

Electronic systems use a number of small devices to
control these circuits, such as
resistors and capacitors, but by far
the most important are transistors.
Transistors can control current in
several ways, but mainly by
amplifying it or switching it on
and off. The secret lies in the
materials they are made from.
These materials, such as silicon
and germanium, have the
remarkable property of conducting
electricity only under certain
conditions — which is why they
are called "semi-conductors."

Valves
*The first electronic components were
thermionic valves, invented in 1904.
Valves like these were used in the
first crude computers in the 1940s.*

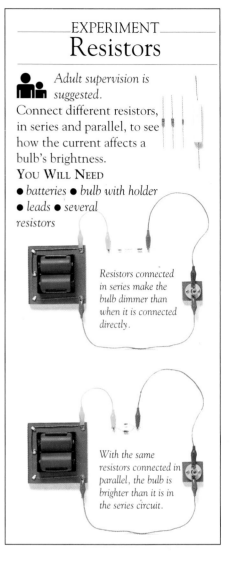

EXPERIMENT
Resistors

*Adult supervision is
suggested.*
Connect different resistors,
in series and parallel, to see
how the current affects a
bulb's brightness.
YOU WILL NEED
● *batteries* ● *bulb with holder*
● *leads* ● *several
resistors*

*Resistors connected
in series make the
bulb dimmer than
when it is connected
directly.*

*With the same
resistors connected in
parallel, the bulb is
brighter than it is in
the series circuit.*

■ Soldering

*Adult supervision
is required for all soldering.*
To make up electronic circuits, you have
to join components by soldering. Solder
is a soft metal that usually comes as
coiled wire. An electrically heated rod
called a soldering iron melts the solder
onto each component. When the solder
cools and solidifies, the joint is made.

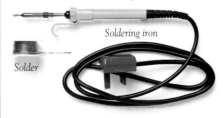

Soldering iron

Solder

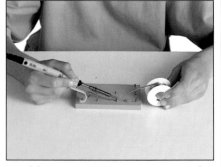

1 SWITCH ON the soldering iron, and,
making sure it is safely supported, let
it heat up. Before each component is
joined, it needs a film of solder melted
onto it. Warm the component wire with
the iron, then place the end of the solder
on the wire and touch the iron tip on the
solder. The solder will melt onto the wire.

2 NOW PLACE THE COMPONENTS so that
the wires that need to be connected
are touching one another. Hold the
end of the solder against the joint, and
heat it with the iron. The solder should
melt in about a second and cool to form
a shiny join. If the joint is gray, melt
the solder again to form a better joint.

■ Transistors

There are two different kinds of semiconductor. In *n* types, current is carried by electrons; in *p* types, it is carried by positively charged holes, which are regions without electrons. Semiconductors are sandwiched together either *p-n-p* or *n-p-n* to make transistors. Each transistor has three connections: the base, the collector, and the emitter. If they are connected up wrongly, the transistor will be ruined. In some, the emitter is marked with a dot. In others, the emitter is shown by a tag on the case.

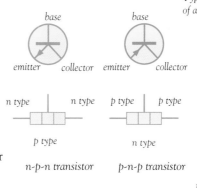

base
emitter   collector

base
emitter   collector

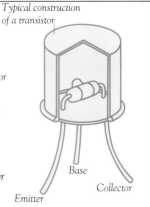

Typical construction of a transistor

Base
Collector
Emitter

n type   *n* type
p type

p type   *p* type
n type

n-p-n transistor   p-n-p transistor

EXPERIMENT

Transistor switch

Adult supervision is suggested for this experiment.

You can use a transistor as a switch to turn a light-emitting diode (LED) on and off. When collector and base are both connected to the battery's positive terminal, the transistor switches the LED on. If you reconnect the circuit by moving the base so that it is connected to the negative terminal of the battery, the transistor remains switched off.

You Will Need
● *n-p-n transistor (BC108)* ● *LED* ● *220-ohm and 10,000-ohm resistors* ● *connecting leads with alligator clips* ● *piece of wood* ● *small nails* ● *short lengths of wire* ● *battery* ● *soldering equipment*

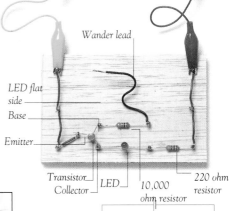

Negative   Positive

Wander lead

LED flat side

Base

Emitter

Transistor
Collector   LED   10,000 ohm resistor   220 ohm resistor

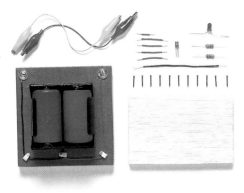

Connecting the base to the positive terminal switches the transistor on, current flows through the LED, which lights up.

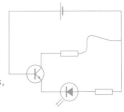

Connecting the base to the negative terminal switches the terminal off. The current stops, and the LED switches off.

Making up the circuit
Push the nails into the wooden base in the pattern shown. Connect the circuit as in the main picture by wrapping the connectors of each component and the wires around the nails, and solder them in

place. Carefully connect the leg on the LED's flat side (the shorter leg) to the collector of the transistor. Turn on the transistor switch by touching the wander lead to the battery's positive side.

A basic radio

IF YOU LISTEN to radio, do you ever wonder how the sound is created? The answer is radiation. Like light, radio signals are waves of electro-magnetic radiation (p. 78). In broadcasting, sounds are changed to an electrical signal with microphones (p. 168). This signal is used to vary the frequency of radio waves transmitted from a radio station. The varying waves re-create the electrical signal in your radio receiver, and this is amplified to reproduce the original sound.

EXPERIMENT
Making a radio

 Adult supervision is required.

A basic radio receiver can be made with just one transistor. If your transistor has four legs, fold back or cut off the one connected to the case.

YOU WILL NEED
● *RF loop-stick antenna* ● *RF choke (inductor)* ● *crystal earpiece* ● *OA79 semiconductor diode or equivalent* ● *AF124 transistor or equivalent* ● *3 resistors & 1 potentiometer*

as below ● *5 fixed & 1 variable capacitors as below* ● *battery & clip* ● *panel pins* ● *balsa wood base* ● *soldering equipment*

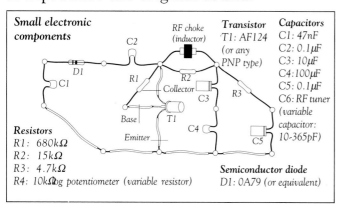

Small electronic components

C2 RF choke (inductor)

D1 C1

R1 R2 Collector C3 R3

Base T1

Emitter C4 C5

Transistor
T1: AF124 (or any PNP type)

Capacitors
C1: 47nF
C2: 0.1µF
C3: 10µF
C4: 100µF
C5: 0.1µF
C6: RF tuner (variable capacitor: 10-365pF)

Resistors
R1: 680kΩ
R2: 15kΩ
R3: 4.7kΩ
R4: 10kΩ log potentiometer (variable resistor)

Semiconductor diode
D1: OA79 (or equivalent)

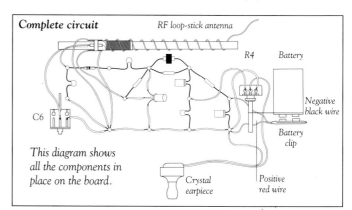

Complete circuit RF loop-stick antenna

R4 Battery

C6

Negative black wire

Battery clip

This diagram shows all the components in place on the board.

Crystal earpiece

Positive red wire

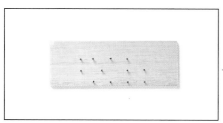

1 PUSH FINISHING NAILS into the balsa wood base in the positions shown in the photograph above. Secure them with glue if necessary.

2 FOLLOWING the diagram above closely, connect all the electronic components on the board, using short lengths of wire where necessary.

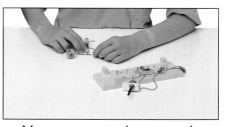

3 MAKE CONNECTIONS by twisting the wires gently together around the nails. Once the wires are contacting properly, glue them in place.

4 IT IS POSSIBLE to make your own radio frequency loop-stick antenna, but it is simpler to buy one. Wind the trailing lead around the rod in a spiral.

5 CONNECT the larger components as shown in the diagram above right. Solder leads to the terminals (see p. 176), without scorching the wood.

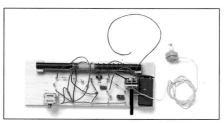

6 ONCE THE COMPONENTS are in place, connect the battery, with the negative wire to the pin between R3 and C5. The radio should now work.

Making a case

The radio will work just as well without a case, but a case makes it look more attractive, and protects it from dust. You can follow the design shown here or come up with your own idea.

● *colored cardboard cut to shapes shown or to your own design* ● *glue* ● *scissors* ● *large cork cut to make two small discs*

1 FOLD OVER THE TABS on the triangular panels. Glue the back panel to the long strip of cardboard that makes the sides of the casing.

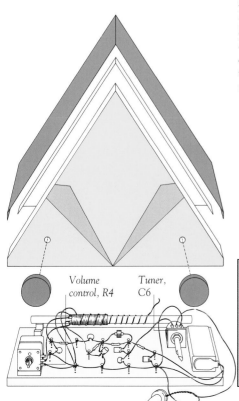

Volume control, R4   Tuner, C6

2 CUT OUT YOUR design motifs, and glue them onto the front panel. Cut small holes corresponding to the tuner (C6) and volume (R4) controls.

4 MAKE SURE ALL connections are secure. Check that the radio is working, adjusting the antenna if necessary. Then fit the sides of the case over the base.

Tuning in

To listen to the radio, connect the battery and put the earpiece in your ear. Turn the volume control (the cork attached to R4) up full, then adjust the tuner (the cork attached to C6) until you hear the radio station you want. This is a very low-powered radio, so you may have to listen quite hard to hear anything. Move the antenna around if necessary to improve the signal.

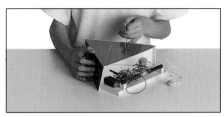

3 THREAD THE LEAD to the earpiece through the hole in the back panel, and connect it to the radio. If it is connected already, cut a slit in the panel.

5 APPLY GLUE to the tabs of the front panel, and slot it into place over the two controls. Glue the cork discs to the controls.

Electronic intelligence

SCIENCE FICTION FILMS and books are full of robots that look, and even think, like humans. Some people believe it will not be long before such machines are a reality. "Intelligent" electronic machines exist today, although most look nothing like people. Such machines cannot yet think in the same way as a person; some say they never will. But they can make decisions and solve problems. An airliner's automatic pilot, for example, can control the plane, even during landing and takeoff — continuously reacting to new electronic data and adjusting the controls accordingly. Computers are the "brains" that give electronic machines their intelligence. Each is programmed with millions of instructions that govern the way they act. Some programs are now so clever that the machine can learn tasks for itself.

Microchips
Tiny complex electronic circuits called microchips lie at the heart of every computer. They store and process information at high speed.

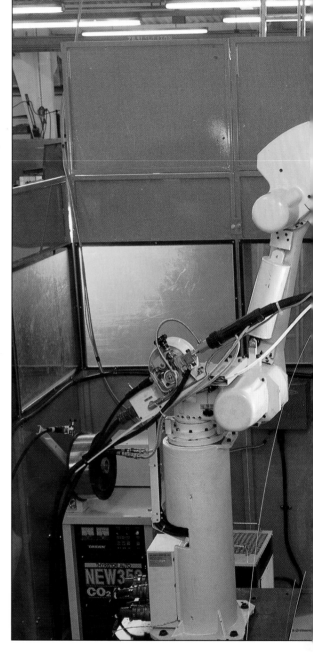

Learning by computer
As computers become more sophisticated and intelligent, they can help people learn. This disc contains a huge amount of information, including film clips, photographs, and text. A computer uses the data on the disc to teach in an "interactive" way. This means that the computer asks a person questions and, depending on the answer it receives, shows different information on the screen.

■ DISCOVERY ■
Babbage's computer

THE HISTORY OF THE COMPUTER goes back to the 17th century, long before the days of electronics. In 1642, French mathematician Blaise Pascal invented the first mechanical adding machine that could add and subtract numbers, using a series of toothed wheels. Almost 200 years passed before the invention of the computer — a machine that could be programmed to do more than one task. This was the *Analytical Engine*, invented in 1834 by Charles Babbage, an English mathematician. Although purely mechanical, Babbage's machine had all the elements of today's electronic computers: a way of giving data and instructions to the machine, a processor for calculations, and a device for printing out answers.

Analytical engine
Babbage spent years working on the design for his computer (above), which consisted of thousands of gear wheels. However, the machine was so complex that Babbage never saw it completed.

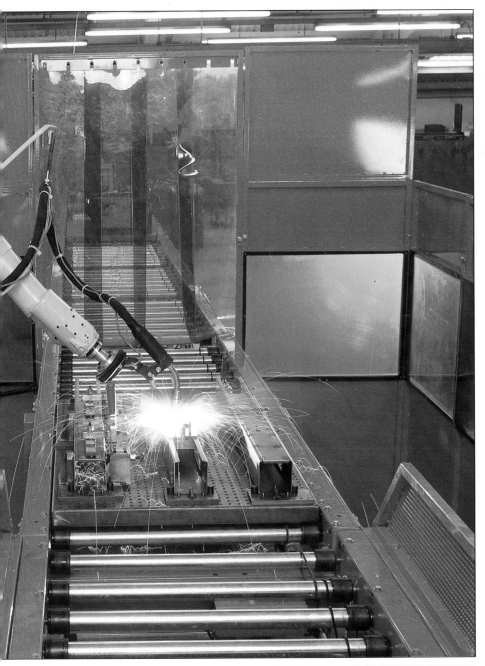

▥ Robots

Many factories rely on robots that paint, weld, and automatically carry components from place to place. Robots never tire of boring or repetitive tasks, and can do many jobs more rapidly and with greater precision than people can. Most factory robots consist of a single arm with a tool such as a spray gun fitted to the end. A skilled worker programs the robot by guiding it carefully through its task; these movements are then stored in the robot's computer memory. The simplest robots merely follow a set of instructions and repeat the same movements again and again. More sophisticated versions have "senses," which they use to pick up information about their surroundings. For instance, the robot welder (left) uses a laser and miniature television cameras to guide itself to the correct position to make a perfect weld. Other robots have touch sensors that help them hold fragile objects without crushing them. Some of the most sophisticated robots are carried by space probes. For instance, the two Viking probes that flew to Mars carried robot explorers that landed on the Martian surface. Equipped with television cameras, automatic arms, and chemical laboratories, the Viking landers analyzed soil samples, photographed the surface, and searched for signs of life.

▥ Computer-Aided Design

Some engineering problems are so complex that they would be virtually impossible without the aid of computers. Computer programs, called computer-aided design systems, help engineers build anything from skyscrapers to microchips. For instance, an engineer designing a bridge needs only to specify a few pieces of information such as the length and width of the bridge. The computer analyzes the forces that the structure will be subjected to and produces a finished design.

Engineers working on the U.S. space program used computer-aided design systems to build the space shuttle.

Glossary

ABSOLUTE ZERO The lowest possible temperature. Zero Kelvin, or –273°C.

ABSORPTION The opposite of radiation — the soaking up of light, heat, sound, and other forms of energy by any substance. Chemical absorption is a solid soaking up a liquid or a liquid or solid soaking up a gas.

ACCELERATION The rate of change of velocity. A body can accelerate by changing either its speed or its direction.

ACID A chemical compound containing hydrogen. Acids are very reactive because in water the hydrogen atoms lose their single electron and become ions. Strong acids (such as sulfuric and nitric acid) are highly corrosive. Mild acids like lemon juice (citric acid) and vinegar (acetic acid) taste sharp.

ACOUSTICS The science of sound. In particular, the way sound is transmitted in buildings, in sound systems, and in musical instruments.

ADDITIVE PRIMARY COLOR The additive primaries are the three basic colors of light — red, green, and blue — that you can mix together to make every other color in the rainbow. They are called "additive" because the other colors are made by adding different amounts of red, green, and blue light. *See also Subtractive primary colors.*

AERODYNAMICS The study of airflow over moving objects – – especially the streamlining of airplanes and cars to cut air resistance or "drag."

AIRFOIL Wing shape that cuts through the air in such a way as to create lift.

AIR PRESSURE The pressure exerted by the movement of molecules in the atmosphere.

ALCHEMY A mixture of basic chemistry, philosophy, and mysticism studied in Europe and the Middle East for more than 1700 years, from the early centuries B.C. into the 17th century A.D. and beyond. Many alchemists sought the "philosopher's stone," the elusive substance that could turn base metals to gold. Others sought an "elixir" (magical potion) that would bring eternal life.

ALKALI A metal or ammonium salt that dissolves in water and neutralizes acids.

ALTERNATING CURRENT (AC) An electric current that continually flows first one way around a circuit and then the other, reversing direction at regular intervals. This is the kind of current put out by the simplest form of generator, an alternator. The electricity supply in most homes is AC.

AMMETER Device for measuring electric currents.

AMPERE Unit of electric current.

AMPLIFIER Electronic device for increasing the strength of an electric signal by drawing energy from a separate circuit.

AMPLITUDE The height of any wave (whether the wave is sound, water, or electromagnetic). In amplitude modulation (AM) broadcasts, the different sounds are carried on the radio wave by tiny changes in the amplitude of the radio wave. In frequency modulation (FM) broadcasts, the frequency is varied.

ANGULAR MOMENTUM The momentum of any spinning object. It is called "angular" because the momentum is always in a straight line, but with a spinning object the angle is constantly changing.

ANIMAL ELECTRICITY In the late 18th century, many people, following Luigi Galvani, believed that electricity was the life force – – the strange force that gave life to flesh and bone.

ANION A negatively charged ion that is drawn to the positive anode in electrolysis.

ANODE The positive electrode in electrolysis or a battery.

Amplification
In a loudspeaker, sound is created by the diaphragm as it vibrates to and fro. The vibrations are controlled by the magnetic coil, which is switched on and off by electric pulses from an amplifier. The amplifier boosts the electrical signal so that the vibrations of the diaphragm are magnified.

Coil

Magnetic

Diaphragm

ANTIMATTER Subatomic particles that are the mirror image of other subatomic particles. The antiparticle of an electron, for example, is a positron. Positron and electron are identical except that the positron has a positive charge and the electron a negative charge.

APERTURE In a camera lens, the size of the opening that controls the image brightness.

ARCHIMEDES' PRINCIPLE When an object floats in water, it pushes out of the way a weight of water equal to its own weight.

ATOM The smallest particle of a chemical element. Particles smaller than an atom have no chemical identity.

ATOMIC NUMBER The number of protons in the nucleus of an atom, which determines which element the atom is.

ATOMIC WEIGHT Because atoms are so small, the mass of atoms of different elements is given in relation to the mass of an atom of an isotope of carbon, called carbon-12, divided by 12.

BAROMETER Device for measuring air pressure.

BASE (CHEMICAL) The opposite of an acid, bases react with acids to form salts.

BASE (ELECTRONIC) In a transistor, the base is the filling in the semiconductor sandwich. It controls the flow of current from the collector to the emitter.

BATTERY A number of cells linked together.

BEAUFORT SCALE Scale of wind-strengths from 1 to 12, devised by naval officer Francis Beaufort in 1806.

BERNOULLI EFFECT When air or water flow past an object, the pressure and velocity of the flow change inversely. So if the object constricts the flow, the pressure drops but the flow speeds up. If the flow broadens out, the pressure rises but the speed drops.

BOILING POINT The maximum temperature a liquid reaches.

BOYLE'S LAW In equal masses of gas at the same temperature, the pressure varies inversely with the volume.

BROWNIAN MOTION The rapid, random, to and fro movement of molecules in a liquid, first observed by Robert Brown in 1827. *See Kinetic theory.*

BUBBLE CHAMBER Device for observing subatomic particles in which a liquid is held under pressure just below its

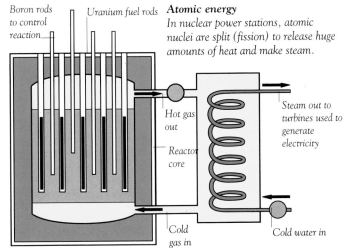

Atomic energy
In nuclear power stations, atomic nuclei are split (fission) to release huge amounts of heat and make steam.

Boron rods to control reaction

Uranium fuel rods

Hot gas out

Reactor core

Cold gas in

Steam out to turbines used to generate electricity

Cold water in

boiling point. When the pressure is suddenly dropped, the liquid boils and tiny bubbles show the paths of the energetic particles.

Calcite Calcium carbonate. Limestone, chalk, and marble are all forms of calcite.
Calcium Very common soft silvery metal usually found as calcite or gypsum.
Capillary action A capillary is a narrow tube. Capillary action is when liquid defies gravity and flows up a narrow tube because of the attraction between the molecules of the liquid and the molecules of the tube.
Carbon Element that occurs in many different forms, including graphite and diamond. Its atomic structure means carbon atoms can form millions of different compounds — so many that organic chemistry is devoted to studying carbon compounds alone. Carbon compounds are the basis of life.
Cathode The negatively charged electrode in a battery, electrolysis, or a tube.
Cathode rays Stream of electrons emitted by the cathode in a glass tube from which nearly all the air has been sucked.

Cathode ray tube The tube behind the screen in TVs and most computers. It is basically a sealed glass tube with a vacuum inside. The stream of electrons emitted by the cathode at one end of the tube makes the fluorescent coating on the inside of the end of the tube glow.
Cation Positively charged ion attracted to the cathode in electrolysis.
Cell Device for generating electricity chemically with two electrodes dipped in an electrolyte. *See Battery.*
Center of gravity The point at which all the mass of an object seems to be focused.
Centrifugal force *See Centripetal force.*
Centripetal force The force that keeps a rotating object moving around in a circle. With a ball whirled on a piece of string, the tension in the string provides the centripetal force. If you let go of the string, the force is lost and the ball flies out. This flying out is called centrifugal force, but it is really a *lack* of force.
Charge The force between atomic and subatomic particles that either attracts them or repels them. Unlike charges attract each other; like charges repel.

Chemical bond The way atoms join together to form molecules. There are three main types: covalent, ionic, and metallic.
Chemical reaction A process in which one or more chemical elements or compounds form new compounds.
Chromatography Technique for separating mixtures of gases or liquids by encouraging the different parts of the mixture to move through an adsorbent material at different rates.
Circuit A loop or loops of electrical conductors that allow current to flow in an unbroken ring from the power source and back.
Cloud chamber Box for observing subatomic particles containing saturated vapor and some liquid. When one wall of the box shoots out, the temperature plummets, and condensation trails show the path of the particles.
Coherer Early device for picking up radio signals consisting of a glass tube containing iron filings held between electrodes.
Combustion Burning; the rapid oxidation of fuel.
Compound Chemical combination of two or more elements.

Concave lens Dished lens.
Condensation Change of gas to drops of liquid.
Conduction (heat) Relay of heat energy through a substance by knocking on of vibrating molecules.
Conduction (electrical) Relay of electrical charge through a circuit by knocking on of free electrons.
Conservation of energy Principle that energy can never be created or destroyed — only converted from one form to another.
Conservation of momentum Principle that momentum is never lost, but simply converted into another form of energy. When a car brakes, for example, the brakes get warm.
Convection When heat is moved through a gas or fluid because heated parts of the gas move themselves. Warm air rises above a heater, for example.
Convex lens Bowed lens.
Corpuscle Old word for particle, used by Newton when talking about a particle theory of light.

Cathode ray tube
The stream of electrons from the cathode scans in lines to and fro across the screen, making the fluorescent coating on the inside of the screen glow.

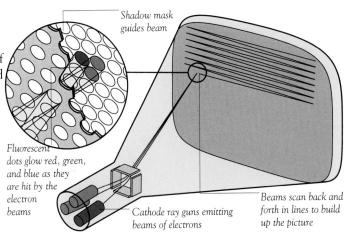

Shadow mask guides beam

Fluorescent dots glow red, green, and blue as they are hit by the electron beams

Cathode ray guns emitting beams of electrons

Beams scan back and forth in lines to build up the picture

CURIE POINT Temperature at which magnetic substances lose their magnetic properties (760°C for iron).

CURRENT (ELECTRIC) Flow of electric charge through a conductor.

DAGUERREOTYPE Earliest successful photographic system, invented by Jacques Louis Daguerre in 1837, using light-sensitive silver salts coated on a metal plate.

DECIBEL Scale usually used for measuring noise levels, but can be used for any kind of power. The scale is logarithmic, which means that doubling the noise level adds 3 to the decibel rating.

DELIQUESCENCE Absorption by a solid of so much moisture from the air that it dissolves.

DENSITY The mass of a substance for a given volume.

DIFFRACTION The bending of waves (sound, light, water, etc.) as they pass an object in their path.

DIFFUSION Gradual mixing of molecules of different substances when they meet because of molecular motion. Gases, in which molecules move very fast, diffuse very quickly. Heating speeds up diffusion.

DIODE Electronic device with two electrodes, used as a rectifier. The earliest diodes were glass valves. Now they are solid semiconductors.

DIP (MAGNETIC) The degree the North pole of a magnet dips toward the ground if allowed to pivot freely.

DIRECT CURRENT (DC) Current that flows only in one direction. *See Alternating current.*

DISTILLATION Purification of liquids by boiling and collecting the vapor as it condenses. The process is

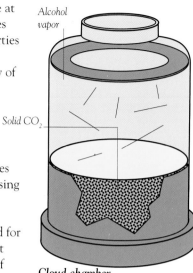

Alcohol vapor

Solid CO$_2$

Cloud chamber
When a radioactive source is placed inside a cloud chamber, the radiation of subatomic particles can be tracked in the alcohol vapor. Solid CO$_2$ in the base keeps it very cold.

used in everything from refining crude oil to making alcoholic spirits.

DISTILLED WATER Tap water is contaminated with a variety of dissolved salts. It can be purified by distillation.

DOPPLER EFFECT The change in the apparent length of a wave (sound, light, or other) as its source moves toward you and away again. You can hear it in the rise and fall in pitch of the noise of a car speeding by. But waves of light from distant stars do the same, producing a slightly redder light (called the "red shift") which enables astronomers to work out how fast stars are moving.

DRAG The slowing effect of air resistance — friction with the air — on a moving object. Liquids can also exert drag.

DYNAMO Electricity generator producing direct current.

ECHO Reflected sound.

ECLIPSE Partial or total blocking out of the Sun from the Moon or Earth.

ELASTIC LIMIT Point beyond which a stretched or squashed substance will not return to its original size.

ELASTICITY The tendency of some materials to return to their original size when squashed or stretched.

ELECTRICITY The effect of charged particles — mostly electrons — on the move (dynamic electricity) or at rest (static electricity). It is one of the most versatile forms of energy.

ELECTRODES The two pieces of metal on each side of a gap in an electrical circuit where it passes briefly through a gas or liquid. *See Anode and Cathode.*

ELECTROLYSIS Chemical reaction caused by passing an electric current through a liquid, usually making the molecules of the liquid break down. Electrolysis of water releases hydrogen and oxygen.

ELECTROLYTE Liquid in electrolysis.

ELECTROMAGNETIC INDUCTION Generation of an electric current in a wire by moving it through a magnetic field.

ELECTROMAGNETIC RADIATION Energy sent out through space by varying electric and magnetic fields. In the 19th century, electromagnetic radiation was thought of as waves. Now it is thought to behave in some ways like a wave and other ways like a particle. *See Photon.* Wave or particle, its speed through the vacuum of space is the fastest possible speed: 299,792,500 meters per

second. Light, radio waves, microwaves, television signals, X-rays, and cosmic rays are all forms of electromagnetic radiation.

ELECTROMOTIVE FORCE Force that makes electric current. *See Potential difference.*

ELECTRON With the proton and neutron, one of the three stable particles in the atom. Electrons are very tiny (about 1/1800 the size of a hydrogen atom) and negatively charged. They normally orbit around the nucleus of the atom, held in place by the opposite charge of the protons in the nucleus.

ELECTROPLATING Coating one metal with a thin layer of another by electrolysis.

ELECTROSCOPE Device for detecting static electric charge, typically using a gold-leaf which lifts according to the size of the charge.

ELEMENT The simplest possible chemicals, each made up entirely of its own particular kind of atom. There are 107 elements altogether, and every substance in the universe is made up from these 107 elements.

ENERGY The ability to do work, from making things move to heating them up. There are many kinds of energy, including electrical energy, nuclear energy,

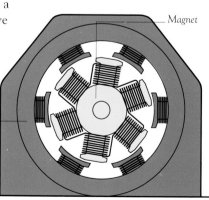

Magnet

Coils

Generator
In power stations, electricity is generated by giant electromagnets rotating within stationary electrical coils.

chemical energy, and heat energy. In everyday life, it is often regarded just as a fuel. For physicists, however, it is one of the basic characteristics of matter, and is interchangeable with mass.

ETHER The mysterious substance philosophers once believed filled the heavens. *See Luminiferous ether.*

ETHERS Volatile, flammable organic chemicals.

EVAPORATION Change of liquid to gas by the escape of molecules from its surface. It can happen at any temperature but speeds up at higher temperatures and lower pressures, or if the surface area of the liquid is increased. It slows down when the air above the liquid approaches saturation.

EXPANSION The gain in size of an object when heated.

FIELD The area over which forces have an effect, including magnetic fields, electric fields, and gravitational fields. The idea of fields of force is fundamental to modern physics, and all matter, from the tiniest subatomic particle to the entire universe, is believed to be held together and moved by four basic types of field.

FISSION (NUCLEAR) Splitting of the nucleus of large atoms (such as uranium) into two or more smaller nuclei to release huge quantities of energy.

FOCUS Point at which the light rays from a lens or group of lenses come together.

FORCE Something that changes shape or motion — stretching or squashing, pushing or pulling.

FORMULA Letters with numbers indicating a particular chemical element or compound. The letters in a formula indicate the type of atom. Thus the letter H is a hydrogen atom; the letter O is an oxygen atom. The suffix numbers indicate the number of those atoms present in the molecule. Thus H_2O, which is the formula for water, shows there are two hydrogen atoms and one oxygen.

FOSSIL FUEL Coal, oil, natural gas, and peat — fuels formed from the remains of trees and

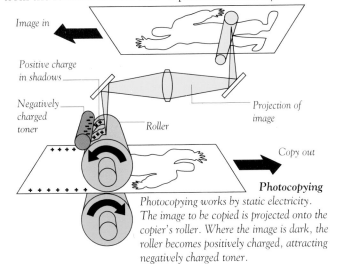

Image in

Positive charge in shadows

Negatively charged toner

Roller

Projection of image

Copy out

Photocopying
Photocopying works by static electricity. The image to be copied is projected onto the copier's roller. Where the image is dark, the roller becomes positively charged, attracting negatively charged toner.

other organic matter long ago and usually buried far underground. Most date from the Carboniferous era, around 300 million years ago.

FREEZING POINT The temperature at which a liquid solidifies. For water this is normally 0°C, but it rises with pressure.

FREQUENCY The number of waves (sound, radio waves, etc.) per second, usually measured in cycles per second or Hertz (Hz). The pitch of a musical note reflects its frequency.

FRICTION The force that resists movement when one surface slides against another.

FULCRUM Point at which a lever pivots.

FUSION (NUCLEAR) Fusing of the nuclei of two or more small atoms such as deuterium to release huge amounts of energy. Nuclear fusion reactions fuel the sun and hydrogen bombs, but have yet to be successfully achieved for power production.

GALVANOMETER Device for measuring small amounts of electric current.

GENERATOR Machine that uses mechanical power to produce electricity.

Generators work by electromagnetic induction — that is, by moving an electric circuit through a magnetic field or vice versa. In the simplest generators, a coil rotates between magnets. In large-scale generators, electromagnets rotate between a series of coils. *See also Dynamo and Alternator.*

GRAVITY Force of attraction between matter, proportional to its mass. Gravity is the force that holds us on Earth and keeps the Earth and the planets orbiting around the Sun. Since all matter has gravitational attraction, you pull the Earth just as the Earth pulls you, but since the Earth's mass is so much bigger, the effect of your pull is negligible.

GREENHOUSE EFFECT The way the Earth's atmosphere prevents much heat from the Sun from escaping into space. Without it, the world would be very cold. But scientists now worry that increases of gases in the air that contribute to the greenhouse effect — such as carbon dioxide from burning fossil fuels — may prevent so much heat from escaping that the world starts to warm up, with potentially disastrous consequences.

GYPSUM Calcium sulfate — e.g., the white stone called alabaster used for statues.

GYROSCOPE Heavy spinning disc pivoted to spin freely at any angle. Because of the disc's angular momentum, it will stay spinning in exactly the same plane — which is why gyroscopes are used in airplanes to indicate whether they are flying level.

HOOKE'S LAW When an object stretches or squashes, the change in size is proportional to the stress.

HUMIDITY The amount of water vapor in the air.

HYDRAULICS Liquids in motion and at rest.

HYDROCARBONS Organic compounds of hydrogen and carbon. There are vast numbers of these, including petroleum and natural gas.

HYDROGEN A colorless, odorless, but highly flammable gas, the simplest and lightest of all elements.

HYDROMETER Simple instrument for measuring the density of a liquid, according to how high it floats.

HYGROMETER Instrument for measuring the humidity (moisture content) of the air.

INCIDENT RAY Ray of light

falling on a mirror, lens, etc. — that is, before reflection or refraction.

INERTIA An object's resistance to change in motion, depending on its mass.

INFRARED Heat radiation. Part of the spectrum of electromagnetic radiation emitted by hot objects.

INFRASOUND Sound too low to be heard by humans.

INSULATION (ELECTRIC) Reduction in the flow of electric current by certain materials (insulators). Also, the covering that contains the current inside a wire.

INSULATION (THERMAL) Reduction in the spread of heat from hot area to cold.

INTEGRATED CIRCUIT Miniature electronic circuit contained within a single crystal of semiconductor — usually a silicon chip.

INTERFERENCE The interaction of two or more waves (sound, light, etc.). When the peaks and troughs of two waves coincide, they reinforce each other. This is called positive interference. Negative interference occurs when peaks in one wave meet troughs in the other (and vice versa), so that they cancel each other out.

ION Atom that has become positively charged by the loss of electrons or negatively charged by the gain of electrons.

ISOTOPES Different forms of the same element with different numbers of neutrons in their atomic nuclei.

KELVIN Standard scientific scale of temperature, corresponding to Celsius but beginning at –273.15°C.

KINETIC ENERGY The energy possessed by a moving object.

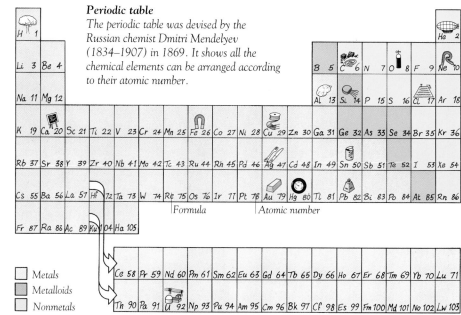

Periodic table
The periodic table was devised by the Russian chemist Dmitri Mendelyev (1834–1907) in 1869. It shows all the chemical elements can be arranged according to their atomic number.

Formula   Atomic number

Metals
Metalloids
Nonmetals

KINETIC THEORY Theory explaining the behavior of solids, liquids, and gases in terms of the movement of the particles (atoms and molecules) from which they are made. A solid melts because heating makes particles vibrate faster and faster, breaking bonds between them. A liquid evaporates because further heating makes the particles vibrate so energetically that they escape from the surface.

LED Light Emitting Diode. Tiny crystals that work like semiconductor diodes but glow when an electric current is passed through them.

LEPTON Believed to be one of the two principal kinds of subatomic particles (with quarks). Electrons are leptons.

LEVER Simple machine with rigid bar that pivots at a fulcrum to make it easier to move a load.

LIGHT The part of the spectrum of electromagnetic radiation that we can see. Newton believed it was streams of particles or corpuscles. In 1801, Thomas

Young showed that light could give interference patterns, indicating that light travels as waves. Now scientists believe that light travels as photons — packets of energy that behave as both particles and waves.

LUMINIFEROUS ETHER When scientists believed light traveled as waves in the 19th century, they assumed that there must be an unknown substance called *luminiferous* (light-bearing) ether to enable the waves to cross the vacuum of space.

MAGNETIC FIELD Area around a magnet in which its effects are felt.

MAGNETISM Invisible force that attracts or repels magnetic materials and has electromagnetic effects.

MASS A measure of an object's inertia — that is, how heavy it is. Contrast with weight, which is a measure of the gravitational force on the object.

MECHANICAL ADVANTAGE When a machine enables you to put less effort in to move a load than you would without

the machine. Specifically, the load divided by the effort.

MELTING POINT Temperature at which a solid becomes a liquid.

MENISCUS Dishing or bowing out of the surface of a liquid due to surface tension.

MOLECULE The smallest naturally occurring particle of a substance, made up from one or more atoms.

NEUTRON Electrically neutral particles in an atom's nucleus.

NITROGEN Colorless gas forming 78 per cent of the atmosphere.

NUCLEUS The dense central core of an atom consisting of protons and neutrons.

OHM Measure of electrical resistance.

ORGANIC CHEMISTRY Study of the chemistry of living matter, consisting entirely of carbon compounds.

OXIDATION Large group of chemical reactions including combustion and corrosion (rusting). Originally thought of as the combining of an element or compound with oxygen (usually from the air). Now applied more generally to a reaction in which the substance loses electrons.

OXYGEN Colorless, odorless gas, making up 21 per cent of the atmosphere. Essential for combustion and for breathing.

PERIODIC TABLE Table of chemical elements ordered by their atomic number.

PHLOGISTON THEORY Mistaken 18th-century theory that every substance that could be burned contained a material called phlogiston that was consumed as the substance burned.

PHOTO-ELECTRIC EFFECT The emission of electrons from a surface when struck by electromagnetic radiation, especially light.

PLASTIC Material that can be molded. Most plastics are man-made polymers.

POLARIZED LIGHT Light in which the waves vibrate in only one direction.

POLYMER Substance made up of very large molecules, often hydrocarbons.

POTENTIAL DIFFERENCE The difference in electrical charge between two places, providing the energy to send an electric current through a circuit.

POWER The rate at which work is done.

PRECESSION The tendency of a spinning body such as a gyroscope to tilt when it is twisted to one side. This is due to the conservation of angular momentum.

PRESSURE The force exerted over a particular area.

PROTONS Large positively charged particles in the nucleus of every atom.

QUANTUM MECHANICS Mathematical system developed in the 1930s and later to explain the motion of subatomic particles.

QUANTUM THEORY Theory developed at the beginning of the 20th century to account for patterns of electromagnetic radiation. It treats radiation as if it is composed of minute packets of energy called *quanta*. (*Quanta* is the plural of *quantum*.)

QUARK With leptons, believed to be one of the two basic kinds of subatomic particle. Protons and neutrons are both made of quarks.

RADIATION Spread of electromagnetic radiation and other subatomic particles.

RADIOACTIVITY The spontaneous breakup of certain nuclei, resulting in the emission of alpha rays (helium nuclei), beta rays (streams of electrons), and gamma rays (very short wave electromagnetic radiation).

RECTIFICATION Conversion of alternating current (AC) to direct current (DC).

REFRACTION Bending of light rays as they pass from one material to another.

RESISTANCE The degree to which a conductor obstructs the flow of an electric current, measured in ohms.

SATURATION Maximum amount something can hold. A saturated solution is one in which the solvent contains as much of the solute as it can hold. When the air reaches saturation point, it can hold no more water vapor, and water will begin to condense.

SCALAR QUANTITY Quantity in which direction is not important, such as temperature and speed.

SCANNING TUNNELING MICROSCOPE Special microscope developed in the 1980s allowing individual atoms to be photographed for the first time.

SEMICONDUCTOR Crystal that conducts electricity only partially. Used in transistors.

SOLUTE Substance dissolved in liquid.

SOLUTION Liquid containing dissolved substance — a solid, a liquid, or a gas.

SOLVENT Liquid in which a substance is dissolved.

SPECIFIC GRAVITY Ratio of the density of a substance to the density of water.

SPECIFIC HEAT Measure of the amount of heat needed to raise the temperature of one kilogram of a substance by one Kelvin.

STRAIN The change in length of an object when stretched, per unit length.

STRESS The stretching or squashing force on an object, per unit area.

SUBATOMIC PARTICLE Tiny particle less than the size of an atom. Hundreds of particles are now known, but most, except for electrons, protons, and neutrons, exist for only a fraction of a second.

SUBTRACTIVE PRIMARY COLORS The primary colors of pigments and paints — yellow, magenta (reddish-purple), and cyan (greenish-blue) — from which all other colors can be mixed. They are called subtractive because they absorb (subtract) other colors from white light. *See also Additive primary colors.*

SURFACE TENSION Molecular force that pulls the surface of any liquid into the minimum possible area, making raindrops round and creating a meniscus in a glass of water.

TERMINAL VELOCITY The maximum velocity that a falling object can achieve.

THERMAL ENERGY Heat energy; the energy of vibrating atoms and molecules.

THERMIONIC VALVE Old electronic component (superseded by transistors) used to rectify or amplify an electrical signal. It was essentially a glass tube containing a metal cathode and metal filament in a gas at very low pressure.

THRUST The force propelling an airplane or rocket.

TRANSISTOR Triode made with a sandwich of semiconductor materials.

TRIODE Electronic component for amplifying, using three electrodes, one of which (the base) controls the strength of the signal passing between the other two (the emitter and the collector).

ULTRASOUND Sound too high in frequency to be heard by the human ear.

ULTRAVIOLET (UV) Electromagnetic radiation with a wavelength just too short for us to see, and making up 5 per cent of the Sun's rays. The ozone layer in the atmosphere protects us and plants from the worst effects of UV radiation, which would otherwise be very dangerous.

VECTOR QUANTITY Quantity with direction, such as velocity.

VELOCITY Speed in a particular direction.

VOLTAGE Measure of electric potential.

WATT Measure of power, usually electrical.

WEIGHT *See Mass.*

WORK Movement or change made by a force.

X-RAY Short-wave electromagnetic radiation to which skin is transparent.

Index

■ ACKNOWLEDGMENTS ■

■ SPECIAL PHOTOGRAPHY
Mike Dunning: pages 1, 2, 6, 16, 34, 81, 82, 83, 92, 93, 94, 95, 104, 105, 106, 108, 109, 110, 111, 112, 113, 118, 120, 121, 123, 148, 149, 150, 151, 152, 153, 155, 156, 157, 162, 163, 164, 166, 167, 168, 169, 170, 171, 176, 177
David Purdy: pages 82, 86, 87
Jerry Young: pages 80–81, 98, 99

■ ILLUSTRATION
David Hallcock: page 110
Nicholas Hewetson: pages 80, 84, 88, 104, 107, 108, 110, 136
Richard Lewis: pages 17, 27, 64, 65, 67, 78, 79, 118, 124, 125, 128, 133, 166, 167, 186
Meridian: pages 95, 120, 121, 123
Rob Shone: pages 92, 93, 96
Richard Ward: pages 100, 101, 102, 103

■ 3-D DRAWINGS
David Hallcock: pages 93, 163
Trevor Lawrence: pages 153, 165, 171
Janos Marffy: pages 49, 50, 68, 178, 179

■ COMPUTER DRAWINGS
Martyn Foote, Dawn Ryddner, Bryn Walls

■ MODEL MAKING
David Donkin: pages 27, 28, 29, 30, 48, 51, 72, 73, 74, 75, 178, 179
David Hallcock: pages 92, 93, 162, 165, 168, 169
James Neville: pages 126, 127
Paul Smith: pages 120, 121
All other models made by Martyn Foote, assisted by Liza Bruml and John Farndon

■ PICTURE CREDITS
t=top; c=centre; b=bottom; r=right; l=left
All Action: 71cl/Duncan Raban
Allsport: 40tr/Bob Martin
Ron Boardman: 176cl, 180bl
British Museum, Geological Museum:158bl
Cavendish Laboratory, Cambridge: 26cl
Bruce Coleman Pictures: 140tr/Gerald Cubitt
Brian Cosgrove: 116t, 116bl, 124tr
Zoe Dominic Pictures: 97tl/Catherine Ashmore
Mary Evans Picture Library: 78tr, 79bc, 81tc, 85br, 88br, 89cr, 89br, 102bc, 103cl, 122bl, 146tl, 150bl
Martyn Foote: 50tl
Ford Motor Company: 58t, 147cr
French Railways: 170tr
Leslie Garland: 68tl
Steve Gorton: 8, 9, 10, 11, 17, 18, 19, 21, 22, 24, 25, 26, 27, 28, 30, 31. 34t, 72, 73
Robert Harding Picture Library: 132tl, 138tr, 141tr
Image Bank: 44cl, 44bc, 45t, 66cl, 79bl, 103cl, 131cr, 131br
Industrial Diamond Review: 17cr
Frank Lane Pictures: 124c
Mansell Collection: 25tr, 44cr, 56tl, 66br, 130cl, 131tc
Diana Morris: 57br
Museum of the Moving Image: 88tl, 92cl
National Gallery, London: 99bl
National Maritime Museum: 79tr
Stephen Oliver: 31b, 37, 38, 39, 40, 41, 51, 52, 53, 54, 55, 59, 60, 61, 62, 63, 64, 65, 68, 69, 70, 71, 72, 73, 74, 75, 81c, 82ct, 83, 84b, 89l, 90, 91, 96, 97, 106l, 107, 122, 125, 126, 127, 129, 132b, 133, 134, 135, 136, 137, 138, 139, 140, 141, 142, 143, 160, 161, 178, 179
Tim Ridley: 32b, 33, 35t, 36b, 119t
Ann Ronan Picture Library: 154tl
Rover Group: 156tr
Science Photo Library: 13, 15tr, 18tr, 24tc, 26bc, 29br, 32tr, 35br, 36tr, 44tl, 45cr, 46tr, 50tr, 52tr, 58bl, 60tr, 62tr, 66tl, 66–67t, 67t, 67tr, 74cl, 76-77, 78tl, 78bl, 79br, 80bl, 100bl, 100–101, 101br, 102cl, 103cr, 114–115, 118cl, 120bl, 121tr, 124tl, 124cc, 124cr, 125cl, 125cr, 130tl, 134tr, 142tr, 143tr, 144–145, 149tr, 152cl, 159tr, 162cl, 164tr, 166tr, 168tr, 178tl, 175tl, 175cl, 180cl, 180-181t, 181tr, 181bc
Tony Stone Picture Library: 42–43
U.S. Navy: 159br
Woodmansterne: 110tl
Zefa: 172–173

■ MANY THANKS TO:
Technical consultant: David Williams, Chief Technician, Acton High School, London

Additional technical advice: Jack Challoner and Ralph Hancock

Additional models: Leona Clarke and Olivia Patt

Editorial, design, and computer help: Lynne Brown, Austin Barlow, Nigel Duffield, Joanna Figg-Latham, Sasha Heseltine, Stephanie Jackson, Jason Little, Nirmala Patel, and Anna Youle

Additional picture research: Melissa Albany, Rehan Ashraf, and Emma Johnson

Radio design: Lionel Wolovitz

Bicycle roller: Stuart Benstead

Bicycles: Sean Thomas, Evans Cycles, The Cut, London SE1

Pigments: Jo Kirby, Scientific Section, National Gallery, London

Model agency: Little Boats, 166 Ifield Road, London SW10

Additional photography: Dean Belcher: 29bl
Andy Crawford: 20b, 23tr, 29b, 36b, 47l, 55c, 69cb, 128bl
Pete Gardner: 60b
Gary Kevin: 30 cb
Tim Ridley: 29cb
Justin Scobie: 8cl
Clive Streeter: 123b
Tim White: 30c
Additional artwork: Chris Lyon: 41br